Quattro Pro® for Scientific and
Engineering Spreadsheets

Robert G. Parks

Quattro Pro® for Scientific and Engineering Spreadsheets

With 91 Figures

Springer-Verlag
New York Berlin Heidelberg London Paris
Tokyo Hong Kong Barcelona Budapest

Robert G. Parks
Sacramento, CA 95864 USA

LaserJet is a trademark of Hewlett-Packard Company
Lotus, 1-2-3 is a trademark of Lotus Development Corp.
Quattro Pro is a trademark of Borland International, Inc.
Postscript is a trademark of Adobe Systems, Inc.
Turbo Basic is a trademark of Borland International, Inc.
Library of Congress Cataloging-in-Publication Data
Parks, Robert G.
    Quattro pro: for scientific and engineering spreadsheets / Robert
G. Parks.
        p.   cm.
    Includes index.
    ISBN-13:978-0-387-97636-5
    1. Engineering—Computer programs.   2. Electronic spreadsheets—
Computer programs.   3. Science—Computer programs.   4. Quattro Pro
(Computer program)   I. Title
TA345.P37   1991
502.85'5369—dc20   91-35103

Printed on acid-free paper.

Production managed by Henry Krell; manufacturing supervised by Robert Paella.
Typeset by Thomson Press (I) Ltd, New Delhi, India.

9 8 7 6 5 4 3 2 1

ISBN-13:978-0-387-97636-5   e-ISBN-13:978-1-4612-2810-3
DOI: 10.1007/978-1-4612-2810-3

Acknowledgments

Appreciation is expressed to Nan Borreson who was the key in converting an idea into reality. The reviews of the mathematics by Don Bristow and the manuscript by Ralph Galantine have provided many valuable contributions to this book.

Robert G. Parks
Sacramento, California
March 31, 1991

Contents

Figures

Tables

Introduction

Before the advent of the digital computer and the electronic calculator, an engineer used a manual spreadsheet called a tabular calculation for the presentation of alternate plans on feasibility studies, physical properties of chemical mixtures and similar problems. The tools were a slide rule, a sharp pencil and a large eraser. It often took hours or days to accomplish what is done electronically now in minutes and even seconds. However, the engineer and scientist still do not have access to many dedicated scientific electronic spreadsheets. They depend for the most part upon those developed for business and accounting purposes. For general applications there are some mathematical packages that will present the data in tabular form and some that will interact with specific spreadsheets.

The electronic spreadsheet is a general-purpose computer program for organizing, analyzing and manipulating data. The advantage of the electronic spreadsheet over the manual tabular method is the speed and accuracy of calculations as well as the ease of introduction of new or modified data. In its present stage of development and availability, the electronic spreadsheet is very much like a programmable calculator with a formidable memory and graphic capabilities. The important feature is the spreadsheet open calculation format that allows another person to review in detail the calculations. Quattro Pro shows the variables and constants on the screen which allows the scientist and engineer to easily follow the calculations. A keyboard switch can replace the data display with the formula used to generate each cell. A formula that has been entered in a spreadsheet cell can be copied either down a column or across a row; Quattro Pro will automatically adjust the data for the new relative positions. Copying not only reduces keystrokes but eliminates the chance of error. The repeated manual entry of a formula into a calculator will always increase the risk of error. An additional advantage of the spreadsheet is that the full size computer printout can contain more information than the calculator paper tape. Trial calculations can be saved as hard copy or to the disk.

The storage register of the programmable calculator and the spreadsheet cell are comparable, as each stores only one number or data item. However, while each cell result of the spreadsheet calculation is retained, some of the programmable calculator intermediate data are usually replaced in the storage registers. If the object of the problem is to provide data pairs, then the retention of all the data is an advantage of the spreadsheet over the calculator.

Quattro Pro was designed for business applications but can be used for scientific calculations that can utilize a spreadsheet format. The functions that are available include trigonometric, natural and base 10 logs, exponential, logical, and statistical functions. Matrix algebra and regression analysis programs are included for more complex calculations. The graphic capabilities are excellent, providing XY graphs from linear to loglog, including hardcopy. Quattro Pro also includes database management capabilities for storing and analyzing data.

Quattro Pro with its low cost and excellent graphics also has the added advantage of being able to export and import comma separated values. External computer programs such as FORTRAN, PASCAL, C, and BASIC can provide data for inclusion in the spreadsheet. The external program can be simple because there is no need to provide arrays. Perhaps the external program would only need to read an import file, manipulate the data, and write it back as an output file.

How to Use This Book

This book is intended to be a supplement to the Borland Quattro Pro *Version 3.0 User's Guide*. A working knowledge of the concepts and procedures of Quattro Pro is assumed. At the beginning of each chapter the Quattro Pro commands used in that chapter have been summarized as a reference. Borland provides an excellent separate and convenient references in the *@ Functions and Macro Guide*.

Only the Quattro Pro procedures and techniques of interest or necessary for engineering and scientific calculations will be emphasized and demonstrated by examples. The examples have been simplified to provide a clear illustration of using the Quattro Pro spreadsheet for engineering and scientific work. It is hoped that other individuals will be encouraged to expand the use of the spreadsheet for their own more sophisticated problems.

Terminology

The units of dimension and their names used in the examples meet the requirements of Le System International d'Unites (SI). Since the United States is converting to the metric system, many of the technical journals and universities now require that all data and calculations be submitted in SI.

The terminology and symbols used in the examples all reflect that background at least to the extent that the special characters are available on the PC system.

Disclaimer

Borland International, Inc. has reviewed the contents of this book prior to final publication. The book does not represent the view, opinions or product strategies of Borland International, Inc. Neither Borland International, Inc. nor Springer-Verlag makes any representations or warranties with respect to the contents of the book. The author advises the readers that the formulas and equations used in the examples in the book are intended solely to illustrate the use of the spreadsheet in scientific and engineering calculations. Under no condition does the author represent or warrant that the contents of the book can be applied for anything other than the instructional purposes for the use of the product described.

1
The Spreadsheet

For the interest of the occasional spreadsheet user, the discussions of Quattro Pro Version 3.0, at the beginning of each chapter will include a summary of the implementing commands. The first chapter summary will include the key operations, precedence of calculations, and functions. Those aspects of Quattro Pro that are of special interest for engineering and scientific calculations will be expanded with examples.

Help

That great engineer, Murphy, noted for his development of the laws of the interactions between people and machines, is said to have made the statement, "When all else fails, read the instructions." Instructions are always at the user's finger tips with the Quattro Pro Help screens. The key for the help screens is F1. The Help screens are quite extensive and are provided with three modes of entry.

Ready Mode

When in the Ready mode, pressing F1 will bring up the Help Topics screen. Each topic has several related screens of information. Further information screens are shown in bold keywords. The related topics are listed at the bottom of the screen. The item wanted for more information can be accessed by highlighting with the arrow keys and pressing ENTER. Pressing ESC returns the spreadsheet to the screen

Menu Mode

If a pulldown menu is activated, and a command given, pressing F1 will bring up Help on that specific command. Keywords and related topics for further information are also provided in the same manner as the Help Topics screen. The keyword or related topic can be highlighted with the arrow keys. Pressing ESC returns the spreadsheet to the screen.

Error Mode

If the unforseen happens and the spreadsheet displays an error message in the middle of the screen, pressing F1 will display an explanation of it. The error messages can also be reviewed by entering the Help Topics screen and choosing Error Messages. Pressing ESC returns the spreadsheet to the screen.

Cell Types

The fundamental spreadsheet unit is the cell, specified by its XY location. The X axis or abscissa is alphabetical and the Y axis or ordinate is numerical such that a cell is identified as an alphanumerical ordered pair. The cell contents can be a value or a label. Quattro Pro checks the first character of the entry and categorizes the entry either as a value or a label.

Value

A value must be either a number or a formula. If the first character of an entry is a value identifier as shown in Table 1.1, then the entry is treated as a value. To use a cell number as a calculation entry requires that the alphabetical first character be preceded by one of the identifiers in Table 1.1. If the entry is a function then the prefix must be the (@) character.

TABLE 1.1. Value identifiers.

```
0 through 9 . + - ( @ # $
```

Numbers

The default alignment of a value in a cell is at the far right. The alignment can be changed to left or center using the /**Style**|**Alignment** command. It should be noted that a mix of left, right and center positions will not affect the calculations. For the sake of clarity, it is best to use a consistant alignment. If the length of a number exceeds the width of the column, the cell display is a row of asterisks (*********). The number will be displayed if the column is widened sufficiently using the command /**Style**|**Width**. The default display for a column is nine digits.

TABLE 1.2. Alternate numeric formats.

```
Fixed.......limits decimal places
Scientific....scientific notation
Currency.......specified currency
,.........................financial
+/- .........horizontal bar graph
Percent...............percentages
Date.................date or time
Text......................formulas
Hidden........suppresses display
```

The formats that are available are shown in Table 1.2. To change the numeric format of a single cell or a block of cells, the command **/Style|Numeric Format** can be used. If a new default format is desired, use the command **/Options|Formats|Numeric** (*Chapter 6, User's Guide*).

Dates and Times

Quattro Pro will attempt to interpret a standard date format as a formula. To signal to Quattro Pro that a date is intended, press CTRL D before entering the date. A complete description of all the other options is available (*Chapter 3, User's Guide*).

Labels

A cell entry of text is designated as a label. The label can include numbers or text used as a description. Quattro Pro requires the label be preceded by an identifier. The identifier can be any letter or punctuation mark other than those shown in Table 1.3. When the cell entry is a label, the typing of the identifier brings up the LABEL mode indicator on the status line. When the label is entered, the status line returns to READY.

TABLE 1.3. Only symbols NOT used as label identifiers.

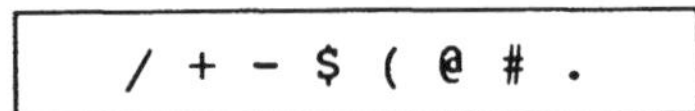

Label Alignment

The default label alignment in Quattro Pro is left justified. As in the case of values, the position can be changed using the command **/Options|Formats|Align Labels** (*Chapter 3, User's Guide*). However, labels can be automatically aligned during entry if the first keystroke is one of the label alignment preface shown in Table 1.4. An example of using a label alignment prefix would be ˆ "quote".

TABLE 1.4. Label alignment prefix.

Label Prefix Character	Alignment
' (apostrophe)	Left Justified
^ (caret)	Center Justified
" (quotation mark)	Right Justified

Numbers as Labels

If a label must begin with a number, as in an address or a sample number, it must be preceded with a label prefix as in Table 1.3. If a prefix is not used, Quattro interprets a number as value or a formula. If the input also contains a letter, then it becomes an illegal entry and Quattro goes into the Edit mode for correction.

Repeating Characters

Often the headings of a spreadsheet are isolated from the data with lines of characters. Repeating characters can be generated across the width of one cell by using the command BACKSLASH (\). For example, typing * will fill a cell with asterisks. The /Edit|Copy command can be used to repeat the characters in the subsequent cells. The |Style|Line Drawing command will also accomplish the same result. If a label beginning with a literal backslash is required, precede the entry with a label alignment prefix.

Formulas

Of most interest in the application of the technical spreadsheet are the formulas. Formulas in Quattro Pro are interpreted algebrically so that the end values are calculated from input values and operators in a standard algebraic order of precedence. Formulas require the use of one of a precedence prefix shown in Table 1.5.

If a formula begins with cell coordinates the prefix + is usually used. For a formula that includes the value of cell A1, the cell A1 would be indentified as +A1. The plus sign will have no effect on the value of A1.

TABLE 1.5. Formula identification prefixes.

0 1 2 3 4 5 6 7 8 9 0 . + − (@ # $

Operators

The result of a formula depends upon the order in which the algebraic operations are performed. The algebraic operators of Quattro Pro and their

TABLE 1.6. Operators.

Operator	Description	Precedence
&	String combination	1
#AND# #OR#	Logical AND, logical OR	1
#NOT#	Logical NOT	2
= <>	Equal, not equal	3
< >	Less than, greater than	3
<=	Less than or equal	3
>=	Greater than or equal	3
- +	Subtraction, addition	4
* /	Multiplication, division	5
- +	Negative, positive	6
^	Exponentiation	7

precedence are shown in Table 1.6. The operations are performed in the order of precedence with the highest precedence being performed first.

The use of parentheses can override the default operator precedence. The liberal use of nested parentheses are encouraged to control the calculation sequence and improve clarity. Quattro Pro gives prompt notification if one of a pair of parentheses is missing. The only indication that a pair of parentheses is missing or misplaced is an obviously wrong answer to the problem.

Types of Formulas

From simple addition to complex calculations, the ability to manipulate cell relationships is the key to number crunching power. The three types of formulas that can be used are arithmetical, textual and logical.

Arithmetic

Arithmetic formulas are algebraic expressions that process numeric data. The @ Functions can be used in the formulas and accept the operators in Table 1.7.

TABLE 1.7. Arithmetic operators.

- + * / ^ = < > <= >= <>

Text

Text formulas are those created with string functions and accept any of the operators in Table 1.8. A detailed explanation of text formulas and the concatenation of strings is available (*Chapter 3, User's Guide*).

TABLE 1.8. Text operators.

&   <   >   <=   >=   =   <>

TABLE 1.9. Boolean operators.

<   >   <=   >=   <>   =   #NOT#   #AND#   #OR#

Logical

Logical formulas are expressions that can be evaluated as true or false. The logical formulas accept the Boolean operators shown in Table 1.9.

Addresses

For the technical user, one of the most important features of the Quattro Pro spreadsheet is the ability to copy formulas by either column or row while controlling the cell references. The default Quattro Pro cell reference is relative. When the formula is copied to another position in the spreadsheet, the copied formula cell addresses are changed relative to the position of the initial cell. If the data of a cell are to be a constant throughout the calculation, the cell address must remain absolute when the formula is copied to another position.

Relative

To copy a formula involving a relative reference, use the command /Edit|Copy. As an example, Figure 1.1 shows a section of a spreadsheet which has data in columns A, B, and C and the formula $A*B/C$ in column D. Using the Copy command with the cursor set on cell D1, the contents of D1 will become the source and the destination will be cells $D2\cdots D4$. The spreadsheet will display the numerical data in each highlighted cell. The formula for each cell will be visible in the upper left corner of the screen. The formula in each of the cells $D2\cdots D4$ is adjusted automatically for the relative reference to the position of the source formula.

	A	B	C	D
1	A1	B1	C1	+A1*B1/C1
2	A2	B2	C2	
3	A3	B3	C3	
4	A4	B4	C4	
5				

FIGURE 1.1. Spreadsheet before using copy.

```
         A            B            C            D
1        A1           B1           C1           +A1*B1/C1
2        A2           B2           C2           +A2*B2/C2
3        A3           B3           C3           +A3*B3/C3
4        A4           B4           C4           +A4*B4/C4
5
```

FIGURE 1.2. Spreadsheet after using copy.

TABLE 1.10. Absolute reference.

```
$B6....fixes column coordinate only
B$6....fixes row coordinate only
$B$6...fixes both coordinates
```

Absolute

The address of a constant can be fixed when a formula is copied. The ($) can be used to fix either the column, row or both as shown in Table 1.10.

Moving a Block

A group of contingent cells is referred to as a block. The contents of a block can be moved with the /**Edit**|**Move** command. The move can be to the same spreadsheet or to a different one, where any existing data at the new location will be overwritten. The default location is the current cell. The cell format as well as the data are moved.

If a block to be moved contains a formula, then care must be exercised, as Move is not the same as Copy. For the Move command, cell references remain the same for both relative and absolute addresses. When a cell containing a formula is moved to another spreadsheet, but the reference cells are not, Quattro Pro links the two spreadsheets. If a referenced cell is moved, Quattro Pro updates the reference formula with the new address.

If data are moved into a cell referenced by a formula, the reference becomes invalid and ERR is displayed. Calculations are stopped until the error is corrected. To reverse a move that has caused an error, use the /**Edit**|**Undo** command (*Chapter 2, User's Guide*).

Recalculation

Quattro Pro recalculates and updates formula results when changes have been made to the reference cell values. Only those formulas that have had refrenced changed cell values are recalculated.

Mode

Three modes of recalculation are available. The default is background mode which enables recalculation to take place between keystrokes. Automatic mode recalculates after data are changed but spreadsheet input pauses until the calculation is finished. The Manual mode only operates when the Calc Key F9 is pressed. The mode of recalculation can be changed by the **/Options|Recalculation|Mode** command.

Order

The command **/Options|Recalculation|Order** is used to set the order of recalculation. The default setting is Natural, which causes each referenced cell to be recalculated before the referenced formula is recalculated. Natural is recommended for accuracy. Before the development of the Natural, spreadsheet calculations were rowwise or columnwise. The row and columnwise calculations are archaic, but are retained in Quattro Pro for special purposes such as circular formulas.

Functions

To the Quattro Pro user who is performing engineering or scientific calculations, the built-in @ Functions are of primary importance. The functions of interest for technical calculations are described here. Borland's documentation gives a complete description of all functions (@ *Functions and Macros, Supplement, User's Guide*).

Mathematical Functions

The general and trigonometric mathematical functions have been separated into tables for ease of reference. The general functions are listed in Table 1.11. The other mathematical functions are trigonometric and are displayed

TABLE 1.11. General mathematical functions.

@ABS(X)	The absolute value of X
@EXP(X)	e raised to the X power
@INT(X)	The integer portion of X
@LN(X)	The Log base e of X
@LOG(X)	The Log base 10 of X
@MOD(X,Y)	The remainder of X/Y
@Pi	The value of Pi
@RAND	A random number between 0 and 1
@ROUND(X,Num)	X rounded to the Num digits (Up to 15)
@SQRT(X)	The positive square root of X

TABLE 1.12. Trigonometric functions.

@Degrees(X)	The number of degrees in X radians
@Radians(X)	the number of radians in X degrees
@SIN(X)	The sine of angle X
@ASIN(X)	The arc sine of X
@COS(X)	The cosine of angle X
@ACOS(X)	The arc cosine of X
@TAN(X)	The tangent of angle X
@ATAN(X)	The arc tangent of X (2 quadrant)
@ATAN2(X,Y)	The arc tangent of Y/X (4 quadrant)

in Table 1.12. The trigonometric functions are regrouped from the alphabetical sequence to one with the function and the arc function adjacent to each other.

Statistical Functions

Quattro Pro provides statistical functions for both spreadsheet and database operations. The mathematical procedures for both are the same and are limited to counting, the measurement of dispersion and linear regression. The analysis of dispersion is left to the user.

Database Statistical Functions

The database statistical functions require that the data be taken from selected field and records of the database. Specific ranges and types of data can be located with a search procedure. Chapter 6 provides a listing of the syntax (Table 6.1) and examples of database operations.

Spreadsheet Statistical Functions

The spreadsheet statistical functions require the data input to be by cell location. The ability to search for selected data is not available. Table 6.2 of Chapter 6 lists the functions. Examples of the use of the spreadsheet statistical functions are given.

Logical Functions

The logical functions in Table 1.13 and the logical operators in Table 1.9 are used to construct conditional statements. The conditional statement returns either the logical value 1 for true or the logical value 0 for false depending upon the evaluation of the expression.

TABLE 1.13. Logical functions.

```
@FALSE                          The logical value 0
@TRUE                           The logical value 1
@IF(Con,TrueExpr,False,Expr)    True Expr if Cond is true,
                                  False Expr if Cond
                                  is false
@ISNUMBER(X)                    1 if X is a numeric value,
                                  otherwise 0
@ISERR(X)                       1 if X is ERR, otherwise 0
@ISNA(X)                        1 if X is NA, otherwise 0
@FILEEXISTS(FileName)           1 if FileName exists,
                                  otherwise 0
@ISSTRING(X)                        1 if X is a string,
                                  otherwise 0
```

Edit

Editing the Quattro Pro spreadsheet is simplified by the built-in cursor movement functions and the cell editing command. If a cell contains only a few characters, deleting the contents and retyping will suffice. But in most engineering and scientific calculations, many of the formulas are too lengthy for retyping to be practical.

F2, the Edit Key

If a formula is to be revised, use the cell selector to highlight the cell and press the Edit key, F2. The cursor and cell contents will then appear on the left part of the status line at the top of the spreadsheet. Changes are made using the DELETE, BACKSPACE, cursor movement and alphanumeric keys. The changes are entered in the usual way with either the ENTER or cursor movement keys.

Cursor Movement Keys

GoTo key, F5, provides a simple and a quick way to move the cell locator to a specified location. Pressing F5 brings up a prompt for a cell address. After typing a valid address, pressing ENTER highlights the cell addressed, redisplaying the screen if necessary.

TABLE 1.14. Cursor movement keys.

Key	Description
Arrow:	
Left	Moves left one cell
Right	Moves right one cell
Up	Moves up one cell
Down	Moves down one cell
Ctrl Left	Moves left one screen
Ctrl Right	Moves right one screen
End Left	Moves* left
End Right	Moves* right
End Up	Moves* up
End Down	Moves* down
PgUp	Moves up one screen
PgDn	Moves down one screen
Home	Moves to A1

*If the current cell has an entry, moves in arrow direction to the next non-blank cell preceding an empty cell. If the current cell is blank, move in arrow direction to next non-blank cell

Files

Unfortunately a computer file is subject to many types of failure. One of Murphy's laws states, 'In computer files, the real life failure is inversely proportional to the calculated probability." To limit the data loss in the event of a file corruption or hardware failure, the best approach is to save the work frequently and back up the hard disk files to removable files on a regular basis.

If several files are open, it is possible to save all of them with the command **/File|Save All**. Each window of the open files is presented, and a request is made to either overwrite or back up the file.

Because user errors in Quattro Pro can cause data loss, it is usually recommended that work files should be saved before any of the following are attempted:

exit Quattro Pro
leave the computer unattended
retrieve a file
copy a block
move a block
combine files
sort data
block fill
extract command

Undo Command

In the event of an error resulting in the possible loss of data, Quattro Pro provides the command **/Edit|Undo or Alt-F5**. The Undo command must be enabled before the mistake occurs, otherwise an error message will be displayed. The Undo command is set by selecting Enable in the **/Options|Other|Undo** menu. To make the selection permanent, choose Update from the Options menu. If the Undo command is enabled, there is a reduction in the speed of program operation. This penalty may be well worth the price if the error would result in a significant recovery time.

Transcript

A Transcript of activity is maintained in the file Quattro.Log. The transcript feature is a maintenance log of all keystrokes since the last Save, Erase or Retrieve command less those still in the keystroke buffer.

The Transcript Window is invoked using the command **/Tools|Macro| Transcript**. The Transcript Menu, Figure 1.3, is opened using the FORWARD SLASH (/) from the transcript window. All the keystrokes since the last Save, Retrieve or Erase command are listed. A vertical line to the left indicates those keystrokes that are in memory but not yet saved to file. All committed keystrokes are recorded in the disk file Quattro.Log.

Undo Last Command

The command history from the most recent list is replayed to the next to the last command.

Restore to Here

The command history from the most recent checkpoint is replayed to the end of the highlighted line in the Transcript Window. The restoration is available for only one file if the work was being done in multiple windows.

```
┌──────────Transcript──────────┐
│                               │
│   Undo Last Command           │
│   Restore to Here             │
│   Playback Block              │
│   Copy Block                  │
│   Begin Block                 │
│   End Block                   │
│   ────────Settings──────────  │
│   Max History Length          │
│   Single Step                 │
│   Failure Protection          │
└───────────────────────────────┘
```

FIGURE 1.3. Transcript menu.

Playback Block

The marked block of actions is played back. The block is marked by highlighting the first line and pressing the command /**Begin Block**. An arrowhead is placed at the beginning line. Highlight the last line and use /**End Block**. Each of the lines in the marked block are indicated with an arrowhead.

Begin Block

Allows the marking of the beginning of a block of actions to be replayed or copied.

End Block

Marks the end of a block of actions.

Max History Length

This command denotes the length of the transcript file before a backup file is made. The .Log file can be set from 1 to 25,000. Setting the .Log file to 0 disables the Transcript function.

Single Step

Playback actions are one keystroke at a time by the command /**Single-Step**. If the Mode is set to yes, the step is manual by pressing the space bar. If Timed is selected, the pause will be for a few seconds.

Failure Protection

This command sets number of keystrokes that are made before the recorded actions are written to the disk. The default is 100 keystrokes. Setting the number to a lower figure slows the program speed because of the increased number of times the recorded actions are written to the disk.

The transcript feature may allow a recovery if the Undo command has not been Enabled. If the error is Erase, then the Undo Last Command will allow Quattro provide a complete recovery. In the event of a power failure, Restore to Here will allow the restoration of work from the last checkpoint to the last recorded action on the disk. Any keystrokes still in the buffer will be lost in power failure.

Managing Files

In addition to the file directory under /**File**|**Open,** there is a second identified as /**File**|**Utility**|**File Manager**. The File Manager is similar to a hard disk directory tree program. Both directories can use the file extension as a filter by changing the extension to *.*. However, the File Manager can operate with any directory or drive and can display the file revision date and number of bytes. If the File Manager is of interest, Borland provides a detailed discussion (*Chapter 10. User's Guide*).

Screen Display Mode

The default display mode of a Quattro Pro spreadsheet is regular text. Regular text is the display of ASCII characters for 80 columns of letters, characters and lines. If the graphics adapter is EGA or VGA, a graphic display B:(WYSIWYG) mode is available for the spreadsheet. The /**Options|Display** menu is used to set the display.

A quick change between the character and graphic mode is provided by Quattro Pro. On the right side of the screen, mouse button number six (WSG) switches to the B:WYSIWYG (What You See Is What You Get) mode. Button number seven (CHR) can be clicked to change to the character mode.

ASCII Text

If an EGA or VGA graphics adapter is not available, then the display setting should be A:80 × 25. The screen will show the ASCII characters in 80 columns by 25 rows. The cursor display is a rectangle in the character mode.

Graphic Spreadsheet Display

With an EGA or a VGA adapter available, the graphics option can be exercised with either B:WYSIWYG or any of the other graphics adapters listed. The selected font is shown on the screen exactly as it can be printed. If the Greek letters or the mathematical symbols (ASCII D224-D255 or HEO-HFF) are used, the A:80 × 25 mode will display them correctly. The B:WYSIWYG mode will display the special or higher ASCII characters *only* with the Hershey fonts (*Chapter 3 and Appendix F, User's Guide*). The WYSIWYG cursor is presented as an arrow.

Printing

The finished spreadsheet, as displayed on the screen or any part of it, can be printed as hardcopy. The spreadsheet can also be saved to disk as an ASCII file that can be imported into some word processing programs. It is then possible to incorporate technical calculations directly into a report. The /**Print menu** of Figure 1.4 will allow the printing of the spreadsheet and provides the commands to print the current graph.

```
┌──────────Print──────────┐
│                          │
│ Block                    │
│ Headings                 │
├──────────────────────────┤
│ Destination              │
│ Layout                   │
│ Format                   │
│ Copies                   │
├──────────────────────────┤
│ Adjust Printer           │
│ Spreadsheet Print        │
│ Print To Fit             │
│ Graph Print              │
│ Quit                     │
└──────────────────────────┘
```

FIGURE 1.4. Print menu.

Block

The /**Print**|**Block** command provides the means of selecting part or all of the spreadsheet to be printed. The block to be printed is either indicated by the point mode or the coordinates typed on the input line.

Headings

The headings menu allows two options, Left Heading and Top Heading. The Headings are rows or columns to be printed on each page of a multipage document. A Left Heading column would allow labels on each page of a spreadsheet wider than a standard paper size. A Top Heading would allow the spreadsheet headings to be printed on each successive page. The Headings are printed in addition to what was included in the /**Print**|**Block** command.

Destination

The spreadsheet printout destination is determined by the command /**Print**|**Destination**.

```
┌─────────Destination─────────┐
├─Draft-Mode Printing─────────┤
│ Printer                     │
│ File                        │
├─Final-Quality Printing──────┤
│ Binary File                 │
│ Graphics Printer            │
│ Screen Preview              │
└─────────────────────────────┘
```

FIGURE 1.5. Destination menu.

Printer

The default printing is made with the ACII text printer option, /**Print**|**Destination**|**Printer**. The spreadsheet must be without special effects, including graphics. To print a spreadsheet as displayed in the B:WYSIWYG mode, the Final Quality option Graphics Printer must be used. If the screen is in the B:WYSIWYG mode but the Graphics Printer is not selected, the printout will be draft quality.

If the screen is not EGA or VGA, the special fonts can be used, but not displayed. The special fonts can only be seen by printing with the Graphics Printer.

File

The spreadsheet is sent to a disk as an ASCII file with the command, /**Print**|**Destination**|**File**. This is a text file that includes all print information. The File command is used to save a spreadsheet for use as text in another program such as word processing. When the file is written to disk, using Quit to exit the menu will append a .PRN extension unless another extension is specified.

Binary File

To save a spreadsheet with final quality graphics to a disk file, use the command/**Print**|**Destination**|**Binary File**. The command is executed with spreadsheet Print. If a file extension is not specified .PRN will be appended.

If a binary file is to be printed from DOS, the DOS COPY command with the /**B** switch must be used.

COPY filename.PRN /B LPT1

Screen Preview

The/**Print**|**Destination**|**Screen Preview** command can be used to view a spreadsheet as it will appear when printed using the graphics printer. The higher ASCII characters, such as Greek letters and mathematical symbols, are available only in the graphic Hershey font and the default draft font.

The Screen Preview Menu appears at the top of the screen. The commands can be actuated with either the mouse or the keyboard. The Screen Previewer Keys are listed in Table 1.15.

The Color command is used to switch to a different color set. A monochrome will be displayed for a color screen system using a monochrome printer.

Previous and Next are used for multiple page reports to move between pages.

A one inch grid is toggled on or off with the Ruler command. The grid can simplify the comparison of the screen and the printed page for modification.

TABLE 1.15. Screen previewer keys.

Key	Effect
ESC	Exits the Screen Previewer
F1	Displays online help
PgUp	Displays previous page
PgDn	Displays the next page
Up key	Scrolls the zoomed display up
Down Key	Scrolls the zoomed display down
Right Key	Scrolls the zoomed display right
Left Key	Scrolls the zoomed display left
Home	Displays the top of the page
End	Displays bottom of page
Enter	Redisplays the zoomed area
Del	Removes the page guide
Ins	Redisplays the page guide

The Guide toggle provides a location outline of the zoomed view on the screen. The outline may be moved with the arrow move keys. By placing the mouse arrow within the outline and holding down the left key, the outline can be moved with the arrow.

Unzoom removes a zoom enlargement.

Zoom provides an 100%, 200% or 400% enlargement of the location outline.

Layout

The default layout conditions are $8\frac{1}{2} \times 11$ inch paper with $\frac{1}{2}$ inch margins and automatic page breaks. The /**Print**|**Layout** menu is used to change the settings for headers, footers, page breaks, margins, dimensions, orientation, and printer codes.

```
┌──────Layout──────┐
│ Header           │
│ Footer           │
├──────────────────┤
│ Break Pages      │
│ Percent Scaling  │
│ Margins          │
│ Dimensions       │
│ Orientation      │
│ Setup String     │
├──────────────────┤
│ Reset            │
│ Update           │
│ Values           │
│ Quit             │
└──────────────────┘
```

FIGURE 1.6. Layout menu.

Headers and/or Footers

If the technical report requires the use of headers, footers or both, use the command /**Print**|**Layout** and select headers or footers. Quattro Pro separates the headers and footers from the page of text. Three lines, minimum, of text are assigned to each header and footer. Two of the lines are used to separate the header or footer from the text. At least one line is used for the text of the header or footer. For a normal page of 66 lines, a one line header and a one line footer will reduce the available lines to 60.

Break Pages

When printing a large spreadsheet of more rows and columns that can fit on a page, Quattro Pro prints the rows of as many columns that will fit the width of a page. The subsequent rows of the original columns are printed on additional pages until the last row of the spreadsheet is reached. Quattro Pro then returns to the first row and moves to the next set of columns and prints the pages to the last row. The pages can be assembled into one large spreadsheet.

Page breaks can be used with the /**Print**|**Layout**|**Break Pages** command and selecting Yes. If no page breaks are required, the Break Pages selection should be No.

If a spreadsheet is oversize for a single page, it is often possible to make it fit a page with the use of compressed font. Another technique is to change the layout from the conventional portrait or vertical to the landscape or length layout.

Percent Scaling

The size of the graphic spreadsheet for both Screen Preview and printing can be changed with the Percent Scaling command. The size range is from 1 to 1000%. The top and left margins of the page remain the same regardless of the spreadsheet size change.

The printing command is Spreadsheet Print. If the Print to Fit command is used, a Percent Scaling command is ignored.

Margins

The Margins menu allows the printing of spreadsheets on wide paper using different fonts. It may be desirable for a spreadsheet with an inserted graph to have the text of both the graph and the spreadsheet printed with the same graphic font. The change of margins may help when deciding the size of the font.

Page Length (1–100) sets the number of lines to be printed on a page. The standard page of 11 inches in length, using a dot-matrix printer that prints 6 lines per inch, gives the default setting of 66. The number of printable lines is affected by top and bottom margins or headers and footers.

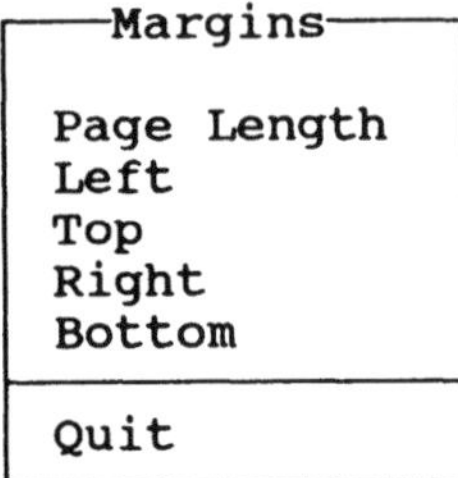

FIGURE 1.7. Margins menu.

Left (0–254) sets the left margin. The default setting is 4 character spaces or approximately $\frac{1}{2}$ inch. If the printed spreadsheet is to be in a bound or looseleaf report, the left margin should be set to accommodate the binding or punch holes.

Top (0–32) determines the space at the top of the report. About $\frac{1}{2}$ inch margin is provided with the default setting of 2.

Right (0–254) sets the right margin. The setting to be used represents the distance from the left edge of the paper to the beginning of the right margin. The default setting is 76, which when used with 10-pitch characters and $8\frac{1}{2}$ inch paper, gives almost a $\frac{1}{2}$ inch right margin.

Bottom (0–32) determines the blank lines at the bottom of the page. The default setting is about $\frac{1}{2}$ inch with the setting of 2.

Dimensions

As noted in the previous sections on margins, the default dimensions used in the layout commands are a character size of 10 characters per inch horizontally and 6 lines per inch vertically. Other options are to use either inches or centimeters as units of dimensions. With the paper size given in inches, in graphics printing it is usually more convenient to set the margins in inches.

Orientation

The default orientation is to print the spreadsheet in the vertical or upright position. The **/Print|Layout|Orientation** menu provides for printing the spreadsheet in a vertical position as Portrait or in a horizontal position as Landscape. The Landscape mode requires that the **/Print|Destination** command must be set for the Graphics Printer.

Setup Strings

The Setup String is a means of providing the codes required by some printers to print special conditions such as compressed or letter quality. Many of the recent printers provide the same options as the Setup Strings by a matrix control panel on the front of the machine. If the use of Setup Strings is

```
┌─────────Format─────────┐
│                        │
│  As Displayed          │
│  Cell-Formulas         │
└────────────────────────┘
```

FIGURE 1.8. Format menu.

necessary, the Quattro Pro documentation has an excellent discussion (*Chapter 4, User's Guide*).

Format

The /**Print**|**Format** menu presents two options for printing a spreadsheet. The default format is the As Displayed option. The spreadsheet will be printed exactly as shown on the screen.

The Cell-Formulas option lists the contents of each cell. The cell contents as shown on the input line are printed one per line. The Cell-Formulas option provides for technical calculations a means of checking the input of complex formulas. When printing in the Cell-Formulas format, Quattro Pro ignores all other print formatting information.

Copies

The number of copies of the spreadsheet to be printed can be set with the /**Print**|**Copies** command. The default number of copies is one.

Adjust Printer

The /**Print**|**Adjust Printer** menu is used to position the paper in the printer. The /**Printer**|**Adjust Printer**|**Skip Line** will move the paper forward one line. To move the printer to the top of the next page the /**Printer**|**Adjust Printer**|**Form Feed** is used. The /**Printer**|**Adjust Printer**|**Align** confirms the paper position and resets the line counter.

Spreadsheet Print

The command /**Print**|**Spreadsheet Print** starts the printing of the spreadsheet to the selected destination.

Print To Fit

If it is desired to print a spreadsheet larger than a standard page, the /**Print**|**Print To Fit** command will proportionally reduce the font and spreadsheet size to accommodate printing on a single page. The limitation on size reduction is the readability of the font. With the print resolution as a limit, Quattro Pro may still require more than one page for a large spreadsheet.

Vapor Pressure

Many chemical engineers were introduced to the use of the tabular calculation or spreadsheet in a sophomore course called Industrial Chemical Calculations. The spreadsheet was used in the calculation of physical properties of chemical mixtures and compounds, material balances and process analysis. One of the first problems was the calculation and plotting of vapor pressures by using methods and data from the literature. One such method is the Antoine equation.

$$\ln P_{vp} = A - B/(T + C) \tag{1}$$

where: P = vapor pressure in mm Hg
          T = temperature in °K
          A, B, C = Antoine vapor pressure coefficients

The vapor pressure in mm Hg would be:

$$P_{vp} = \exp(A - B/(T + C)) \tag{2}$$

Normally, the Antoine equation would be a subroutine in a computer program called up as a result of an change in temperature. However, a picture of how the vapour pressure varies with temperature can be seen by plotting the log of the vapour pressure vs. the temperature. The method of calculation is straightforward. The semi-log plot is made with the logarithmic graphic capabilities of Quattro Pro.

The following example demonstrates the calculation for the alcohols of carbon chain length from 1 to 4. The Antoine coefficients are:

Compound	ANT A	ANT B	ANT C	Temp range (°K)
Methanol	18.59	3626.55	−34.29	257–364
Ethanol	18.92	3803.98	−41.68	270–369
i-Propanol	17.55	3166.38	−80.15	285–400
n-Butanol	17.22	3137.02	−94.43	288–404

Reference: Reid, Prausnitz and Sherwood, "The Properties of Gases and Liquids," Appendix A, McGraw-Hill, 1977.

The spreadsheet is set up using the title and data shown in Figure 1.9. The header row across the top of the columns can be produced by using an underline (__), asterisk (*) or other figure in the first cell and then copying it from column A through L. The titles of the columns are shown in Figure 1.9.

	A	B	C	D	E	F
1						
2	VAPOR PRESSURE USING ANTOINE COEFFICIENTS					
3						
4		LnPvp = A−B/(T+C)				
5		where: P = mmHg				
6		P = exp(A−B/(T+C)				
7	COMPOUND	ANT A	ANT B	ANT C		
8	Methanol	18.59	3626.55	−34.29		
9	Ethanol	18.92	3803.98	−41.68		
10	Propanol	17.55	3166.38	−80.15		
11	Butanol	17.22	3137.02	−94.43		
12						
13	---					
14	TEMP	TEMP	METHANOL	ETHANOL	PROPANOL	BUTANOL
15	°C	°K	mmHg	mmHg	mmHg	mmHg
16	---					
17	10	283	55	23	7	2
18	20	293	97	44	14	4
19	30	303	163	79	28	9
20	40	313	265	134	52	18
21	50	323	415	221	91	33
22	60	333	632	352	153	59
23	70	343	937	542	246	99
24	80	353	1354	813	382	162
25	90	363	1915	1190	576	255
26	100	373	2652	1701	844	387

FIGURE 1.9. Vapor pressure spreadsheet.

Column B is the calculation of the temperature in °K. Each of the alcohol vapor pressures is calculated for the corresponding temperature in °K. A sample calculation for methanol vapor pressure at 10°C using equation (2) and the cell locations of the spreadsheet for the Antoine coefficients is:

$$mmHg = @\,EXP(\$C\$8 - \$D\$8/(B17 + \$E\$8)$$

It should be noted that if the Copy command is used to generate the formulas for the rest of the temperatures, Quattro Pro by default adjusts the formula for a relative reference. To use a constant such as the Antoine coefficients requires that the cell position for each calculation be fixed using the absolute reference \$. Figure 1.11 is a complete listing of all the cell formulas including the key strokes. The apostrophe (') is used to denote a label. The caret (^) will denote and center a label in a cell. Note the differences in the formulas of the text and those as labels in the spreadsheet. Quattro Pro does not support the use of (^) subscripts or superscripts in labels, however the special ASCII characters can be used.

The semi-log vapor pressure graph of Figure 1.10 illustrates the capabilities of Quattro Pro. Chapter 2 outlines the procedures for obtaining presentable graphs.

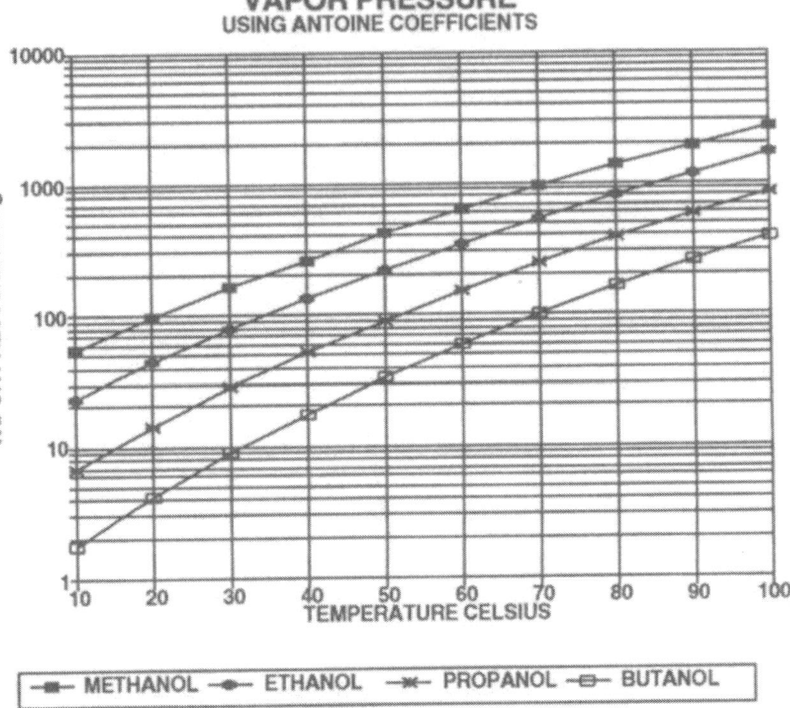

FIGURE 1.10. Alcohol vapor pressures.

```
A2:  '      VAPOR PRESSURE USING ANTOINE COEFFICIENTS
C4:  'Ln Pvp = A - B/(T+C)
C5:  'P = mmHg
E5:  'T = °R
B7:  'COMPOUND
C7:  ^ANTA
D7:  ^ANTB
E7:  ^ANTC
B8:  'METHANOL
C8:  18.59
D8:  3626.55
E8:  -34.29
B9:  'ETHANOL
C9:  18.92
D9:  3803.98
E9:  -41.68
B10: 'PROPANOL
C10: 17.55
D10: 3166.38
E10: -80.15
B11: 'BUTANOL
C11: 17.22
D11: 3137.02
```

(*Continued*)

```
E11:  -94.43
A14:  ^TEMP
B14:  ^TEMP
C14:  'METHANOL
D14:  ^ETHANOL
E14:  ^PROPANOL
F14:  ^BUTANOL
A15:  ^^°C
B15:  ^^°K
C15:  ^mmHg
D15:  ^mmHg
E15:  ^mmHg
F15:  ^mmHg
A17:  10
B17:  273+A17
C17:  (F0) @EXP($C$8-$D$8/(B17+$E$8))
D17:  (F0) @EXP($C$9-$D$9/(B17+$E$9))
E17:  (F0) @EXP($C$10-$D$10/(B17+$E$10))
F17:  (F0) @EXP($C$11-$D$11/(B17+$E$11))
A18:  20
B18:  273+A18
C18:  (F0) @EXP($C$8-$D$8/(B18+$E$8))
D18:  (F0) @EXP($C$9-$D$9/(B18+$E$9))
E18:  (F0) @EXP($C$10-$D$10/(B18+$E$10))
F18:  (F0) @EXP($C$11-$D$11/(B18+$E$11))
A19:  30
B19:  273+A19
C19:  (F0) @EXP($C$8-$D$8/(B19+$E$8))
D19:  (F0) @EXP($C$9-$D$9/(B19+$E$9))
E19:  (F0) @EXP($C$10-$D$10/(B19+$E$10))
F19:  (F0) @EXP($C$11-$D$11/(B19+$E$11))
A20:  40
B20:  273+A20
C20:  (F0) @EXP($C$8-$D$8/(B20+$E$8))
D20:  (F0) @EXP($C$9-$D$9/(B20+$E$9))
E20:  (F0) @EXP($C$10-$D$10/(B20+$E$10))
F20:  (F0) @EXP($C$11-$D$11/(B20+$E$11))
A21:  50
B21:  273+A21
C21:  (F0) @EXP($C$8-$D$8/(B21+$E$8))
D21:  (F0) @EXP($C$9-$D$9/(B21+$E$9))
E21:  (F0) @EXP($C$10-$D$10/(B21+$E$10))
F21:  (F0) @EXP($C$11-$D$11/(B21+$E$11))
A22:  60
B22:  273+A22
C22:  (F0) @EXP($C$8-$D$8/(B22+$E$8))
D22:  (F0) @EXP($C$9-$D$9/(B22+$E$9))
E22:  (F0) @EXP($C$10-$D$10/(B22+$E$10))
F22:  (F0) @EXP($C$11-$D$11/(B22+$E$11))
A23:  70
B23:  273+A23
C23:  (F0) @EXP($C$8-$D$8/(B23+$E$8))
D23:  (F0) @EXP($C$9-$D$9/(B23+$E$9))
E23:  (F0) @EXP($C$10-$D$10/(B23+$E$10))
F23:  (F0) @EXP($C$11-$D$11/(B23+$E$11))
```

(*Continued*)

```
A24:  80
B24:  273+A24
C24:  (F0)  @EXP($C$8-$D$8/(B24+$E$8))
D24:  (F0)  @EXP($C$9-$D$9/(B24+$E$9))
E24:  (F0)  @EXP($C$10-$D$10/(B24+$E$10))
F24:  (F0)  @EXP($C$11-$D$11/(B24+$E$11))
A25:  90
B25:  273+A25
C25:  (F0)  @EXP($C$8-$D$8/(B25+$E$8))
D25:  (F0)  @EXP($C$9-$D$9/(B25+$E$9))
E25:  (F0)  @EXP($C$10-$D$10/(B25+$E$10))
F25:  (F0)  @EXP($C$11-$D$11/(B25+$E$11))
A26:  100
B26:  273+A26
C26:  (F0)  @EXP($C$8-$D$8/(B26+$E$8))
D26:  (F0)  @EXP($C$9-$D$9/(B26+$E$9))
E26:  (F0)  @EXP($C$10-$D$10/(B26+$E$10))
F26:  (F0)  @EXP($C$11-$D$11/(B26+$E$11))
```

FIGURE 1.11. Vapor pressure formulas.

Differential Equations (Runge-Kutta Method)

One of the most powerful mathematical tools used to describe physical events is the differential equation. In most of the engineering equations, the conditions are set by an initial value such that the solution is a particular function that meets a set of specific values. In addition, most of the differential equations encountered cannot be solved in the classical sense, but can only be approximated using numerical methods. The computer has provided relief from tedious calculations and encouraged the development of more sophisticated and accurate methods. The Quattro Pro spreadsheet lends itself to the calculation of the ordered pairs of the solution.

A differential equation that can be classically solved will be used to compare the true solution values with the approximations of Quattro Pro.

$$dy/dx = 2xy \qquad y(1) = 1 \tag{1}$$

by separating the variables:

$$\int y/dy = 2 \int xdx \tag{2}$$

integrating and combining constants:

$$\ln_e y = x^2 - c \tag{3}$$

substituting the initial values of $x = 1$, $y = 1$,

$$y = \exp(x^2 - 1). \tag{4}$$

The approximate solution to the above differential equation can also be

	A	B	C	D	E	F
1						
2		SOLUTION OF A DIFFERENTIAL EQUATION				
3		4th Order Runge-Kutta Method				
4	f(y) = 2xy		y(1) = 1		h =	0.1000
5						
6	--					
7	x	y	F(1)	F(2)	F(3)	F(4)
8	--					
9	1.0000	1.0000	0.2000	0.2310	0.2343	0.2715
10	1.1000	1.2337	0.2714	0.3150	0.3200	0.3729
11	1.2000	1.5527	0.3726	0.4348	0.4425	0.5188
12	1.3000	1.9937	0.5184	0.6083	0.6204	0.7319
13	1.4000	2.6116	0.7313	0.8634	0.8826	1.0483
14	1.5000	3.4902				

FIGURE 1.12. Runge–Kutta test solution.

made by numerical methods. While not as precise as the classical method, the approximation may be the only answer if it not possible to determine an explicit or implicit formula. Most of the numerical computational algorithms use the method of tangent lines and one is known as the secant method. Some methods require the solution for the next order derivative.

One of the more accurate and popular numerical procedures for initial value differential equations is the 4th Order Runge-Kutta method. Using the Runge-Kutta method, it is possible to achieve the accuracy of the Taylor Series expansion without computing higher-order derivatives and in agreement out to the fifth term. The constants F_2 and F_3 in the formula are approximations to the slope at the midpoint of the interval between x_n and x_{n+1}. The location of the midpoint is expressed by $x_n + h$, where h is the incremental step.

The 4th Order Runge-Kutta equation is:

$$y_{n+1} = y_n + (F_1 + 2F_2 + 2F_3 + F_4)/6 \tag{5}$$

where the constants are:

$$F_1 = hf(x_n, y_n) \tag{6}$$

$$F_2 = hf(x_n + h/2, y_n + F_1/2) \tag{7}$$

$$F_3 = hf(x_n + h/2, y_n + F_2/2) \tag{8}$$

$$F_4 = hf(x_n + h, y_n + F_3) \tag{9}$$

The constants are calculated using the increments in x and y as shown in the equations (6–9) for the function. The initial values are used to determine F_1. For F_2, x is increased by h/2, y is increased by $F_1/2$, and the function is recalculated. Each of the subsequent constants is determined and the formula calculated. The solution to the example problem is shown in Figure 1.12.

BEGIN FIGURE

```
F4:  0.1
A9:  1
B9:  1
C9:  +$F$4*(2*A9*B9)
D9:  +$F$4*2*((A9+0.5*$F$4)*(B9+0.5*C9))
E9:  +$F$4*2*((A9+0.5*$F$4)*(B9+0.5*D9))
F9:  +$F$4*2*(A9+$F$4)*(B9+E9)
        o                    o                    o
        o                    o                    o
        o                    o                    o
A13: 1.4
B13: +B12+(C12+2*D12+2*E12+F12)/6
C13: +$F$4*(2*A13*B13)
D13: +$F$4*2*((A13+0.5*$F$4)*(B13+0.5*C13))
E13: +$F$4*2*((A13+0.5*$F$4)*(B13+0.5*D13))
F13: +$F$4*2*(A13+$F$4)*(B13+E13)
A14: 1.5
B14: +B13+(C13+2*D13+2*E13+F13)/6
```

FIGURE 1.13. Test spreadsheet formula printout.

COMPARISON
APPROXIMATION vs ACTUAL

x	y	$\exp(x^2-1)$
1.0000	1.0000	1.0000
1.1000	1.2337	1.2337
1.2000	1.5527	1.5527
1.3000	1.9937	1.9937
1.4000	2.6116	2.6117
1.5000	3.4902	3.4903

FIGURE 1.14. Accuracy test comparison.

The formula for each cell is shown in Figure 1.13. It should be noted that the value of h is an absolute reference and is designated by the $F$4 notation. Once the formulas for line 8 have been entered, lines 9 through 13 are automatically adjusted for relative reference by Quattro Pro using the Copy command.

Figure 1.14 indicates that the difference between the approximation and the absolute value to four places is negligible. The percent difference is $+0.0038\%$ for x = 1.4000 and $+0.0029\%$ for x = 1.5000.

Differential Equations (Sliding Motion)

A more practical problem would be to determine the time necessary to stop a 6.8 kilogram sliding mechanism on a packing line. The initial velocity is 3.75 m/sec and the weight is subjected to a forward force of $1.8(1-t)$ kilograms, where t is time in seconds. A damping force of $0.9082e^{0.4072v}$ is applied where v is velocity in m/sec, and is derived in the chapter on linear regression (Chapter 4). The equation of motion is:

$$(w/g)(dv/dt) + 0.9082e^{0.4072v} = 1.8(1-t).$$

Inverting and solving for dt/dv:

$$dt/dv = -w/g[4(1-t) - 0.9082e^{0.4072v}]$$

The spreadsheet can be set up by setting time equal to 0 at an initial velocity of 3.75 m/s. The velocity is then decreased in 0.25 m/s steps to zero. The Runge-Kutta equation can now be solved for each step by calculating the constants that will provide the time for the next step. With $h = 0.25$ m/s, Figure 1.15 shows the spreadsheet set up.

	A	B	C	D	E	F
1						
2		SOLUTION FOR A SLIDING MECHANISM				
3		4th Order Runge-Kutta Method				
4						
5		dt/dv = -W/g(1.8(1-t) - 0.9082exp(0.4072v))				
6	W=	6.8000 Kg		g=	9.8067 m/s^2	
7	h=	0.2500				
8						
9	v	t	F(1)	F(2)	F(3)	F(4)
10						
11	3.7500	0.0000	0.0728	0.0781	0.0780	0.0839
12	3.5000	0.0781	0.0819	0.0886	0.0884	0.0960
13	3.2500	0.1668	0.0907	0.0991	0.0987	0.1082
14	3.0000	0.2659	0.0985	0.1085	0.1081	0.1197
15	2.7500	0.3744	0.1046	0.1162	0.1157	0.1293
16	2.5000	0.4907	0.1086	0.1214	0.1208	0.1361
17	2.2500	0.6122	0.1103	0.1239	0.1234	0.1399
18	2.0000	0.7363	0.1100	0.1241	0.1235	0.1407
19	1.7500	0.8607	0.1083	0.1223	0.1218	0.1392
20	1.5000	0.9833	0.1055	0.1192	0.1188	0.1359
21	1.2500	1.1029	0.1022	0.1154	0.1150	0.1315
22	1.0000	1.2186	0.0986	0.1111	0.1108	0.1264
23	0.7500	1.3301	0.0949	0.1066	0.1064	0.1211
24	0.5000	1.4371	0.0912	0.1022	0.1021	0.1158
25	0.2500	1.5397	0.0877	0.0979	0.0978	0.1106
26	0.0000	1.6380				

FIGURE 1.15. Sliding mechanism solution.

```
B5:  'dt/dv = -hW/g(1.8(1-t) - 0.9082exp(0.4072v))
A11: 3.75
B11: 0
C11:-$B$7*$B$6/($E$6*(1.8*(1-B11)-0.9082*@ EXP(0.4072
     *A11)))
D11:-$B$7*$B$6/($E$6*(1.8*(1-B11+0.5*$B$7)-0.9082*@ EXP
     (0.4072*(A11+0.5*C11))))
E11:-$B$7*$B$6/($E$6*(1.8*(1-B11+0.5*$B$7)-0.9082*@ EXP
     (0.4072*(A11+0.5*D11))))
F11:-$B$7*$B$6/($E$6*(1.8*(1-B11+$B$7)-0.9082*@ EXP
     (0.4072*(A11+E11))))
              o                 o                 o
              o                 o                 o
              o                 o                 o
A25: 0.25
B25:+B24+(C24+2*D24+2*E24+F24)/6
C25:-$B$7*$B$6/($E$6*(1.8*(1-B25)-0.9082*@ EXP(0.4072
     *A25)))
D25:-$B$7*$B$6/($E$6*(1.8*(1-B25+0.5*$B$7)-0.9082*@ EXP
     (0.4072*(A25+0.5*C25))))
E25:-$B$7*$B$6/($E$6*(1.8*(1-B25+0.5*$B$7)-0.9082*@ EXP
     (0.4072*(A25+0.5*D25))))
F25:-$B$7*$B$6/($E$6*(1.8*(1-B25+$B$7)-0.9082*@ EXP
     (0.4072*(A25+E25))))  A26: 0
B26: +B25+(C25+2*D25+2*E25+F25)/6
```

FIGURE 1.16. Sliding mechanism formula printout.

The formulas in Figure 1.16, while appearing quite complex, are nothing more than the Runge-Kutta substitutions for x and y in the differential equation used to calculate the constants. The first row is copied by Quattro Pro for the rest of the spreadsheet, automatically adjusting each cell's relative values and keeping the absolute values. Note that in the interest of conserving space only the first and last rows are shown in Figure 1.16.

2
Graphs

The text and numeric information contained in a spreadsheet is provided in detail. By contrast a graph will present the same text and data as an overall view and more clearly display the trends of the variables.

Ten different types of graphs are available with Quattro Pro. The majority of the graphs are designed for business purposes. While the bar and pie graphs will find occasional use for engineering and scientific reports, the XY graph for data pairs will normally be more relevent.

This chapter will cover the generation of the XY graph with Quattro Pro. Changing the appearance of a graph with grids, titles, legends and other advanced features will be reviewed. Options such as inserting a graph into a spreadsheet, and naming and saving a graph will be described.

XY Graph

In the /**Options**|**Display**|**A:80** × **25** mode, the Graph menu, Figure 2.1, provides the entry to the data, text and specialized menus. If the B:WYSIWYG mode is chosen, the graph menu is made up of icons illustrating the type of graph. In the case of determining whether the graph is to have lines, symbols or both, the menu may be five layers deep.

The Quattro Pro default graph type is a stacked bar. The discussion of the graph for technical calculations will be limited to the XY graph. The command to select an XY graph type is /**Graph**|**Graph Type**|**XY**.

Series

The term series is used by Quattro Pro to specify the values from the spreadsheet to be plotted on the graph. Up to six series or sets of values can be assigned to the ordinate or Y-axis. The ordinate values must be referenced to a single range of data for the abscissa or X-axis. Six sets of data pairs may be plotted on each graph, provided that they all have the same X-axis values in common.

```
Graph Type
Series
Text

Customize Series
X-Axis
Y-Axis
Overall

Insert
Hide
Name
View                        F10
Fast Graph                      Ctrl-G
Annotate
Quit
```

FIGURE 2.1. Main graph menu.

The Series menu of Figure 2.2 is called up using the /**Graph|Series** command. The first through the sixth series are assigned to the ordinate or Y-axis. Each series is entered by highlighting the series number and pressing ENTER. The block identifying the column is shown on the command line and opposite the series on the menu. The X-Axis series is used to provide scaling information. The command /**Graph|Series|Group** allows the specification of a block of values to be plotted. The block will be divided into one series per column.

When plotting data pairs, each series *must* contain the same number of values. A label cell in a block will receive a value of 0. An empty cell in a block series will wreak havoc with the plot. Once the X-series and the Y-series have been selected, the graph can be presented on the screen using /**Graph|View or F10**.

```
1st Series
2nd Series
3rd Series
4th Series
5th Series
6th Series
X-Axis Series
Group
Quit
```

FIGURE 2.2. Series menu.

```
1st Line
2nd Line
X-Title
Y-Title
Secondary Y-Axis
Legends
Font
Quit
```

FIGURE 2.3. Text menu.

Text

If the graph is to be used in a presentation, the titles can be added with
/**Graph**|**Text**. A main title and subtitle can be used, as well as a title for
each axis and the legend. The font, size, and the color of the graph should
be determined.

As in the Series Menu, the highlighted item when Enter is pressed allows
the text to be typed on the Command line. When ENTER is pressed after
typing, the text is shown opposite the item. If a cell entry is to be used as a
title, typing BACKSLASH (\) followed by the cell address will enter the cell
address on the item line. The cell contents will show a title on the graph.

First Line

The First line or the main title is automatically centered above the graph.
The maximum number of letters is 39. If the title length approaches the
maximum, the font size will be adjusted to compensate for the length.

Second Line

The second line of the main title is automatically centered and is of smaller
type than the first line. The limit on the number of letters is 39.

X-Title

The X-Title is centered horizontally below the X-axis. The length cannot
exceed 39 characters. The font can be different than the main titles, but must
be the same as the Y-Title.

Y-Title

The position of the Y-Title is to the far left of the Y-axis and is read from
bottom to the top. The text is limited to 39 characters and the font must be
the same as the X-Title.

Secondary Y-Axis

Additional text can be displayed vertically to the right of a second Y-axis.

```
1st Line
2nd Line
X-Title
Y-Title
Legends
Data & Tick Labels
Quit
```

FIGURE 2.4. Font menu.

Legends

If more than one Y-Series is to be plotted, each series must have some form of identification. The Legends menu is under /**Graph**|**Text**|**Legends** and allows identification of up to six ordinate series. The Legend text is limitd to 19 characters for each series. A cell entry can be used as a legend by typing BACKSLASH(\) and the cell address. The menu will show the cell address, not the label; the label will be displayed on the graph.

The default position of the legends is to the right of the graph. The alternate position is at the bottom of the graph. /**Graph**|**Text**|**Legend**|**Position** will allow locating the legend, or if none is selected the legend will be deleted.

Fonts

There are eleven basic fonts available for graphs. The default font is Bitstream Swiss. If a graph is intended for presentation it may be advantageous to have the fonts similar to those in the rest of the report. As with the other Bitstream fonts, the higher ASCII characters are not supported.

When each of the items in Figure 2.3 is highlighted and ENTER is pressed, the characteristics shown in Figure 2.4 are available for selection.

Requesting the Typeface option of Figure 2.4 will lead to a list of available typefaces. The current typeface in use is highlighted.

Appendix F, User's Guide provides a description of each of the fonts. Changing the font of a graph text requires the /**Graph**|**Text**|**Font** command. After selecting the text to be changed, choose Typeface and the Available

```
Typeface          Bitstream Swiss
Point Size        36 Point
Style
Color             Blue
Quit
```

FIGURE 2.5. Fonts option menu.

```
Bitstream Dutch
Bitstream Swiss
Bitstream Courier
Roman
Roman Light
Sans Serif
Sans Serif Light
Script
Old English
Eurostyle
```

FIGURE 2.6. Available fonts menu.

Fonts when the menu of Figure 2.6 is displayed. The graph with the new typeface can be sent to the screen with F10.

The internal fonts of Postscript and Laser jet are supported by Quattro Pro. If the installed printer is either of these printers, the Available Fonts Menu, Figure 2.6, will show the fonts available.

The initial display of the main title is 36 points, the subtitle is 24 points and all other text is 18 points. A point is the typography unit measurement of 1/72nd inch. The text can be sized from 6 to72 points. The fonts of the Laserjet printer are preset and cannot be changed in size.

To change the font size, return to the Font Options menu, Figure 2.5, and choose Point Size. After changing the size, the effect can be viewed with F10.

The normal font is the default style for graph text. The typeface can be changed by the **/Graph|Text|Font** command with the Style menu of Figure 2.7. The Style command is limited. The Bold Style is only for Bitstream Dutch and Bitstream Swiss fonts. The Italic Style can only be used with Bitstream Dutch, Bitstream Swiss, Bitstream Courier and Roman fonts. Quattro Pro will accept the command for other fonts but the command will be ignored upon printing.

The font of a title or text can be returned to regular print. The text is located with **/Graph|Text|Font** and the change is accomplished with **Style|Reset**.

The command **/Graph|Text|Font|Color** can be used to change the color of the graph text. Changing to the graphics display mode will present a gallary of actual colors on a color monitor.

```
┌─────Style──────┐
│                │
│  Bold          │
│  Italic        │
│  Underlined    │
│  Drop Shadow   │
│  Reset         │
│  Quit          │
└────────────────┘
```

FIGURE 2.7. Style menu.

```
┌──Customize Series──┐
│                    │
│  Colors            │
│  Fill Patterns     │
│  Markers & Lines   │
│  Bar Width         │
├────────────────────┤
│  Interior Labels   │
│  Override Type     │
│  Y-Axis            │
│  Pies              │
│  Update            │
│  Reset             │
│  Quit              │
└────────────────────┘
```

FIGURE 2.8. Customize series menu.

Customize Series

The customizing of a graph may make a better presentation but can also be used to make a graph more legible. The identification of each ordinate series by line and marker is a necessity.

Colors

Each series is assigned a different color by default. These colors can be changed. By using the command /**Graph**|**Customize Series**|**Colors** a menu will be displayed showing the colors of each series. A pictorial display of colors will be shown if the screen is set to graphics with the /**Options**|**Display Mode**. To make the changed colors the new default colors, use the /**Graph**|**Customize Series**|**Update**.

Fill Pattern

The Fill Pattern paints the bar graphs and pie charts segments with the same selected pattern. There are 14 different fill patterns available.

Markers and Lines

The identification of the six different possible lines and data points is made with lines and markers. Individual markers for the data points and the type of line for each series can be selected. The command to customize the markers and lines is /**Graph**|**Customize Series**|**Markers and Lines** of the menu of Figure 2.9.

Eight different types of lines are used by Quattro Pro to identify between series in an XY graph. The default line styles for the current graph are highlighted.

The Line Style menu of Figure 2.10 lists the lines available. By setting the screen display to B:WYSIWYG mode with the mouse button or the /**Options**|**Display** Mode, a pictorial gallery of line styles is displayed.

```
┌──────Markers & Lines──┐
│                       │
│  Line Styles          │
│  Markers              │
│  Formats              │
│  Quit                 │
└───────────────────────┘
```

FIGURE 2.9. Markers and lines menu.

```
┌──────Line Styles──┐
│                   │
│  Solid            │
│  Dotted           │
│  Center-line      │
│  Dashed           │
│  Heavy Solid      │
│  Heavy Dotted     │
│  Heavy Centered   │
│  Heavy Dashed     │
└───────────────────┘
```

FIGURE 2.10. Line styles menu.

```
┌──────────Markers──────────┐
│                           │
│  A - Filled Square        │
│  B - Plus                 │
│  C - Asterisk             │
│  D - Empty Square         │
│  E - X                    │
│  F - Filled Triangle      │
│  G - Hourglass            │
│  H - Square with X        │
│  I - Vertical Line        │
│  J - Horizontal Line      │
└───────────────────────────┘
```

FIGURE 2.11. Markers menu.

If the markers that have been assigned to each series are not suitable, /**Graph**|**Custom Series**|**Markers and Lines**|**Markers** will bring up the Marker menu to allow a change. The marker that is assigned to the series is highlighted. If the screen display mode is set to B:WYSIWYG Mode, a pictorial representation of the marker symbols is shown. If the changed marker symbols are to be the new default symbols for the program, use the /**Graph**|**Customize Series**|**Update** command.

The /**Graph**|**Customize Series**|**Markers and Lines**|**Formats** command will determine how the lines and/or markers are displayed for an XY graph. Lines or Symbols will allow either lines or markers to be shown. The command Both will set up the use of lines and markers with the graph. The command Neither removes both lines and markers from the graph which in effect deletes a series.

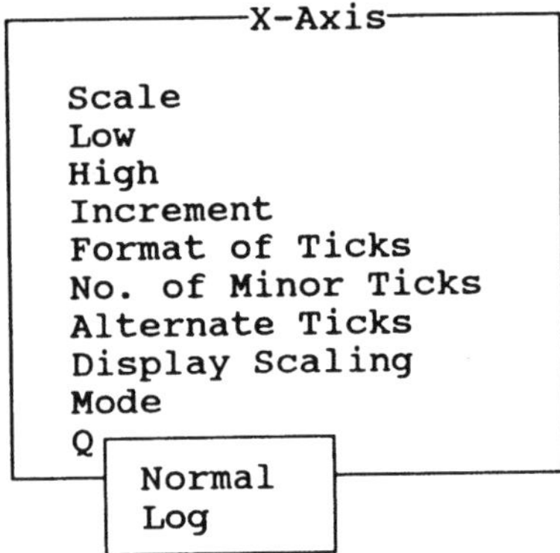

FIGURE 2.12. Formats menu.

FIGURE 2.13. X- or Y-axis menu.

X- or Y-Axis

The default axis scale condition is automatic. Quattro Pro adjusts the scale of the axes to best fit the assigned range of numbers. To reset the scale of the X-axis or the Y-axis of an XY graph, /**Graph**|**X-Axis** or |**Graph**|**Y-Axis** will bring up the X-Axis or Y-Axis menus for change.

Scale

The Scale options are Automatic or Manual. The selection of a manual scale brings a request for the number to use for the low end of the scale. The default number is 0. If all the block values are to appear on the graph, the Low axis number should be less than the lowest graph value. The high end of the scale should be similarly treated; the High axis number should exceed the highest block number.

Increment

The Increment request for a tick mark interval value is limited to line graphs. The command No. of Minor Ticks will provide similar results for the XY graphs.

Format of Ticks

The Format of the Ticks uses the Alternate Numeric Format of Table 1.2 to set the display format of the numeric tick labels.

No. of Minor Ticks

If the numbers of the labels on the X-axis and Y-axis overlap, some of the labels can be replaced with minor tick marks by the command No. of Minor Ticks.

Alternate Ticks

To improve clarity, the Alternate Ticks command will display the ticks using two levels of the text and alternate spacing.

Display Scaling

If the values of the series are large numbers, the Display Scaling will automatically divide the tick labels by 1,000 or 10,000 and indicate the multiplier next to the axis. This feature can be turned off by choosing NO.

Mode

The Mode menu allows the scaling of the X-axis and/or the Y-axis as linear or logarithmic. The term Normal in the Mode menu is logically linear.

Overall

The command /**Graph**|**Overall** provides for the customizing of the entire graph. This feature includes grids, color, and outlines.

Grid

The Grid command sets the position, color and style of the graph grid lines. The grids provide a reference for visual evaluation and interpretation of the data. The position of the grid lines can be referenced to the X-axis only as Horizontal, and the Y-axis only as Vertical or Both axes. Clear will remove the grid from the graph.

```
┌────────Overall────────┐
│                       │
│  Grid                 │
│  Outlines             │
│  Background Color     │
│  Three-D              │
│  Color/B&W            │
│  Drop Shadow Color    │
└───────────────────────┘
```

FIGURE 2.14. Overall menu.

```
┌──────Grid──────┐
│                │
│  Horizontal    │
│  Vertical      │
│  Both          │
│  Clear         │
├────────────────┤
│  Grid color    │
│  Line Style    │
│  Fill Color    │
│  Quit          │
└────────────────┘
```

FIGURE 2.15. Grid menu.

The Grid Color command displays a color menu. If the screen is set to B:WYSIWYG or the /**Options**|**Display Mode**, the display is a visual presentation of the colors available. The Grid Color command controls the color of the outlines of the main title, legends, and the overall graph.

The Line Style menu for the grids is identical to the Line Style menu of Figure 2.10. If the screen is set to the **B:WYSIWYG** mode with /**Options**|**Display Command** or the mouse button the text menu is replaced with a pictorial display.

The Fill Color sets the color behind the grid lines.

Insert

The /**Graph**|**Insert** command will insert a named or the current graph into the spreadsheet. If the monitor is the Enhanced Graphics Adapter (EGA) or the Video Graphics Array (VGA), and the screen is set to the graphics mode, the inserted graph is viewed with the spreadsheet. If the monitor is not EGA or VGA then the graph area is only highlighted and the inserted graph can only be seen by printing.

After naming a graph or selecting the current graph, Quattro Pro prompts for the block to insert the graph. The graph is drawn in the given block designated by pointing or coordinates.

The graph is drawn in the assigned block with a 4:3 aspect ratio. If the graph block has a different ratio, the adjustment can be made in the Layout menu of the printing command. It should be noted that the graphic mode applies to the entire graph and spreadsheet and only a graphic font will be used for all numbers and text. If the same font is to be used for the entire page, not only the font for the graph but the font for the spreadsheet must be selected.

Again it should be noted that the Bitstream fonts do not include the higher ASCII characters. If the Bitstream fonts are to be used, then substitution must be made for the Greek letters and mathematical symbols.

```
┌──────Name──────┐
Display
Create
Autosave Edit
Erase
Reset
Slide
Graph Copy
```

FIGURE 2.16. Graph name menu.

Hide

A graph can be removed from the spreadsheet by using the command /**Graph**|**Hide**. The name of the graph is requested by a prompt. When the graph is deleted, the blank cells of the spreadsheet are shown.

Name

Since Quattro Pro can only save the current graph with the /**File**|**Save** or the Save As command, the /**Graph**|**Name** menu is used to store more than one graph with a file. By assigning names, more than one graph can be stored with a spreadsheet. All graphs can then be retrieved from within the spreadsheet and inserted in the spreadsheet.

Display

The Display command replaces the current graph of a spreadsheet with a previously named graph. /**Graph**|**Name**|**Display** displays a list of existing named graphs and requests the graph name for retrieval.

Create

If another graph in addition to the current one is to be generated from the spreadsheet, the current graph can be saved for recall by naming it with /**Graph**|**Name**|**Create**. The Create command displays the list of existing named graphs, and requests the name. The data currently in the spreadsheet will reflect the newly named graph.

Erase

A graph can be deleted with the command /**Graph**|**Name**|**Erase**. The graph name can either be typed in or selected from the list.

Autosave Edits

With the /**Graph**|**Name**|**Autosave Edits** toggle set to Yes, the current graph is automatically saved before another graph is displayed. If the toggle is set to No, changes since the last Create will be lost.

```
┌──────Graph Print──────┐
│                       │
│ Destination           │
│ Layout                │
│ Go                    │
│ Write Graph File      │
│ Name                  │
│ Quit                  │
└───────────────────────┘
```

FIGURE 2.17. Graph print menu.

Reset

/**Graph**|**Name**|**Reset** is a destruct command. All the saved graphs will be deleted if the answer to the confirmation menu is Yes.

View

The View or F10 command appears in several menus. View allows a quick visual check of the development of the graph or as changes are made.

Fast Graph

The /**Graph**|**Fast Graph** command is limited to the X-axis with labels and cannot be used for a standard XY graph.

Annotate

The /**Graph**|**Annotate** command provides a built-in drawing program. Most corporations, universities and technical publications have specific rules on the formatting of graphs for technical reports. The acceptance of technical calculations and graphics will usually depend more on the content than the particular graphic design.

Printing

After the completion of a graph from the spreadsheet, Quattro Pro can provide a hardcopy for inclusion in a file or report. The graph can be printed either as a separate page using a portrait or landscape format or as an insertion into the spreadsheet.

Destination

The /**Print**|**Graph Print**|**Destination** menu is used to determine where the printout of the graph will go. The setup of the Printer with the /**Options**|**Hardware**|**Printer**|**Default Printers** command determines how Quattro will print, file or preview the screen.

File

Printing the file to disk saves the graph for printing at a future time. The binary file is stored as if the file were going to the graphics printer.

The DOS command to print the binary file is:

COPY filename.PRN /B LPT1

The switch **/B** sends the binary file to printer port LPT1.

Graphics Printer

The designated graphics printer must be capable of handling bit mapped graphics. The majority of nine and twenty-four pin dot matrix printers are supported with driver programs.

A PostScript printer can be specified as the default printer. The PostScript printer will allow a wide selection of fonts. The chosen PostScript fonts will not be visible on the screen, but are replaced with a stand-in font. With the exception of the high ASCII characters, the selected PostScript fonts then are used by Quattro Pro to print the graph.

Screen Preview

With the exception of PostScript fonts, the completed graph can be viewed on the screen as it will appear when printed with the default printer. The use of the /**Print|Graph Print|Destination|Screen** Preview will display the graph complete with formatting, fonts and orientation. The Guide command on the menu bar to set the page guide for the zoom mode. The page guide shows an outline of what part of the graph is being displayed. The zoomed area outline can be moved with arrow keys or a mouse.

Layout

Often in a technical report, the graph of the calculations is confined to a single page as an insert or addition. On the other hand, the graph may be

```
┌─────────Layout─────────┐
│                        │
│  Left Edge             │
│  Top Edge              │
│  Height                │
│  Width                 │
│                        │
├────────────────────────┤
│  Dimensions            │
│  Orientation           │
│  4:3 Aspect            │
├────────────────────────┤
│  Reset                 │
│  Update                │
│  Quit                  │
└────────────────────────┘
```

FIGURE 2.18. Graph layout menu.

required to be pasted as part of the text. Quattro Pro allows freedom to design the graph layout to suit all conditions.

The default margins and dimensions for graphs by Quattro Pro are measured in inches; the default dimensions for the spreadsheet are in characters. The use of inches or centimeters is recommended for graphics printing. It should be noted that all layout values should be non-zero for all the values to be effective.

The default aspect ratio of Quattro Pro for a graph is 4:3, where the height is three quaters of the width. With the command **/Print|Graph Print|Layout|4:3 Aspect|Yes** the default ratio is in effect. No matter what the size set for the graph by the Height and Width dimensions, the width of the graph fits the selected Width and the height is adjusted to meet the 4:3 ratio. Changing the 4:3 Aspect to No will allow the aspect ratio to fit directly any set of assigned dimensions.

Orientation

As with the spreadsheet, the graph can be printed in a vertical portrait position or a horizontal landscape position. The landscape orientation can be used to completely fill a page for insertion into a technical report being prepared with a word processing program. If the spreadsheet and the graph can be accommodated on a single page, the command **/Graph|Insert** can be used. On the other hand, if the spreadsheet occupies most or more than a page, the graph can be oriented as portrait and juxtapositioned with the text into a reserved space with the word processing program.

Inserting a Graph in a Spreadsheet

As an example of inserting a graph into a spreadsheet, the Vapor Pressure example of the previous chapter will be used. It is worth repeating that in order to display the combined spreadsheet and graph on the screen, an EGA or VGA display adapter is required. If a monochrome or CGA screen is used, the inserted graph appears as highlighting only. In either case, the **/Options|Display Mode** must be set to B:WYSIWYG.

Developing the Graph

First the main graph menu **/Graph** as shown in Figure 2.1 is accessed. The following Table 2.1 is developed from the main graph menu. Each command is taken in sequence and the graph parameters are displayed.

The graph can be saved with the spreadsheet. The screen display is changed to Graphics with **/Options|Display**. To insert the graph, use the **/Graph|Insert command**. The location of the graph block will be at the bottom of the page, A27···F42. The graph as shown in Figure 2.2 was set with the 4:3 ratio for the assigned space.

The Hersey font Sans Serif was used for the entire page including the spreadsheet. The spreadsheet of Figure 2.19 is a copy of Figure 1.9. If a Bitstream font had been used, the temperature column headings of °C and °K would have had to be shown as deg C and deg K. The font could have been selected to match an available font from a word processing program. Unless the word processing program can support a binary file, the page must be pasted into the report.

Fourier Series

Periodic functions are a part of many engineering problems, such as mechanical vibration, alternating currents, and engine torque. The simplest form of a periodic function is the sinusoid, $y = Y \sin 0$. The sinusoidal form of driving force and steady state solution is used when possible. Many of the periodic phenomena are not as simple as the sine wave, and require the use of the Fourier Series for solution.

A mechanical vibration problem can usually be modeled as a force acting on a weight suspended by a spring, with the motion restricted by a snubbing or damping device. An example would be Figure 2.21. If the data in Table 2.2

	A	B	C	D	E	F
1						
2	VAPOR PRESSURE USING ANTOINE COEFFICIENTS					
3						
4		LnPvp = A-B/(T+C)				
5		where: P = mmHg				
6		P = exp(A-B/(T+C)				
7	COMPOUND	ANT A	ANT B	ANT C		
8	Methanol	18.59	3626.55	-34.29		
9	Ethanol	18.92	3803.98	-41.68		
10	Propanol	17.55	3166.38	-80.15		
11	Butanol	17.22	3137.02	-94.43		
12						
13	--					
14	TEMP	TEMP	METHANOL	ETHANOL	PROPANOL	BUTANOL
15	°C	°K	mmHg	mmHg	mmHg	mmHg
16	--					
17	10	283	55	23	7	2
18	20	293	97	44	14	4
19	30	303	163	79	28	9
20	40	313	265	134	52	18
21	50	323	415	221	91	33
22	60	333	632	352	153	59
23	70	343	937	542	246	99
24	80	353	1354	813	382	162
25	90	363	1915	1190	576	255
26	100	373	2652	1701	844	387

FIGURE 2.19. Spreadsheet for graph insertion.

VAPOR PRESSURE USING ANTOINE COEFFICIENTS

$$LnPvp = A-B/(T+C)$$
where: $P = mmHg$
$$P = exp(A-B/(T+C))$$

COMPOUND	ANT A	ANT B	ANT C
Methanol	18.59	3626.55	-34.29
Ethanol	18.92	3803.98	-41.68
Propanol	17.55	3166.38	-80.15
Butanol	17.22	3137.02	-94.43

Temp °C	Temp °K	Methanol mmHg	Ethanol mmHg	Propanol mmHg	Butanol mmHg
10	283	55	23	7	2
20	293	97	44	14	4
30	303	163	79	28	9
40	313	265	134	52	18
50	323	415	221	91	33
60	333	632	352	153	59
70	343	937	542	246	99
80	353	1354	813	382	162
90	363	1915	1190	576	255
100	373	2652	1701	844	387

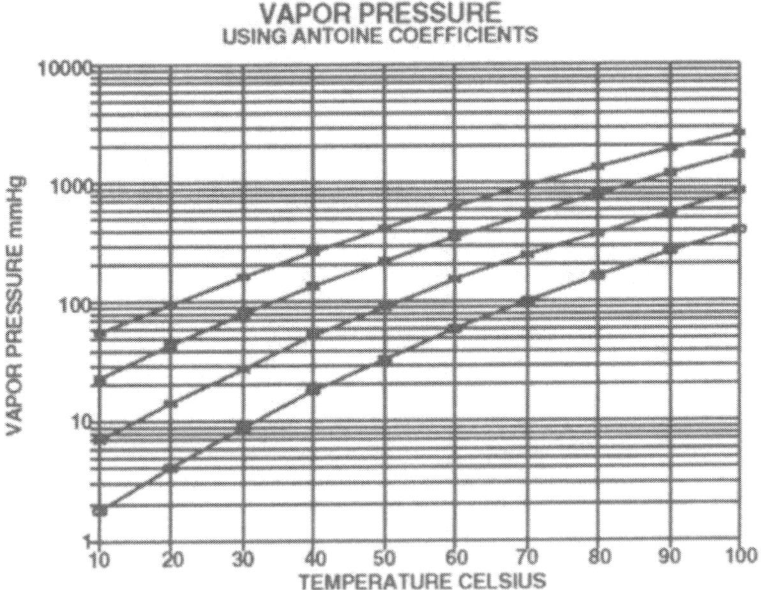

FIGURE 2.20. Inserted graph.

TABLE 2.1. Insert graph parameters.

```
|Graph Type                                              XY

|Series
   1st Series                                       C17..C26
   2nd Series                                       D17..D26
   3rd Series                                       E17..E26
   4th Series                                       F17..F26
   X-Axis Series                                    A17..A26

|Text
   1st Line                                   VAPOR PRESSURE
   2nd Line                    USING ANTOINE COEFFICIENTS
   X-Title                          TEMPERATURE CELSIUS
   Y-Title                          VAPOR PRESSURE, mmHg
   Legends
      1st Series                                  METHANOL
      2nd Series                                   ETHANOL
      3rd Series                                  PROPANOL
      4th Series                                   BUTANOL
   Font                             (Hershey) Sans Serif

|Customize Series
   Markers & Lines
      Lines Styles                                   Solid
      Markers
         1st Series                        A-Filled Square
         2nd Series                                 B-Plus
         3rd Series                             C-Asterisk
         4th Series                        D-Empty Square
      Format                                          Both

|X-Axis                                           Automatic
   Mode                                              Normal

|Y-Axis                                              Manual
   Low                                                    1
   High                                               10000
   Mode                                                 LOG

|Overall
   Grid                                                Both
```

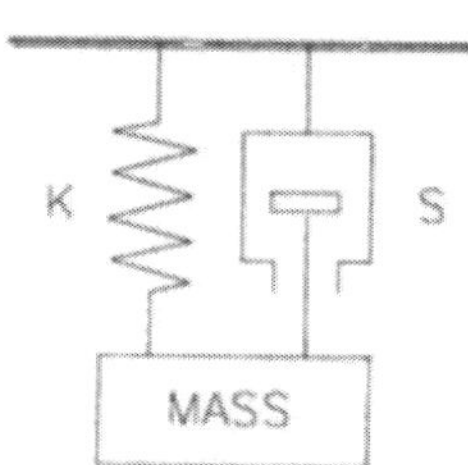

FIGURE 2.21. Vibration model.

TABLE 2.2. Vibration data.

Movement of the spring support	Y = 0.0011 meters
Movement Frequency	f = 0.636 Hertz
	n = 4.0000 radians/s
Mass attached to the spring	M = 0.2978 Newton
Spring unit stretch	K = 42.8890 kg/m
Snubbing force	S = 0.4460 kg-sec/m

	A	B	C	D	E	F
1			FOURIER SERIES			
2						
3			Data Input			
4		Y =	0.0011 m			
5			n =	4.0000 rads/s		
6			f =	0.6360 Hz		
7			M =	0.2978 N		
8			K =	42.8890 kg/m		
9			S =	0.4468 kg-sec/m		
10						
11						
12			Amplitude and Phase Angle of Odd Harmonics			
13	---	---	---	---	---	---
14	Harmonic		1	3	5	
15	---	---	---	---	---	---
16	Phase Angle,rads		0.0468	1.5697	3.2583	
17	Phase Angle,deg		2.6840	89.9380	186.6857	
18	Amplitude,m		0.0014	0.0033	0.0001	
19						
20			Calculation of Harmonics			
21	---	---	---	---	---	---

	Time	Funda-mental	Harmonics 1st	Harmonics 3rd	Harmonics 5th	Total Motion
24	---	---	---	---	---	---
25	0.00	0.0000	-0.0001	-0.0033	0.0000	-0.0034
26	0.01	-0.0011	-0.0000	-0.0033	-0.0000	-0.0033
27	0.10	-0.0011	0.0005	-0.0012	-0.0001	-0.0008
28	0.20	-0.0011	0.0010	0.0024	0.0001	0.0035
29	0.30	-0.0011	0.0013	0.0030	0.0001	0.0043
30	0.40	-0.0011	0.0014	-0.0003	-0.0001	0.0010
31	0.50	-0.0011	0.0013	-0.0032	0.0001	-0.0018
32	0.60	-0.0011	0.0010	-0.0020	0.0001	-0.0009
33	0.70	-0.0011	0.0005	0.0017	-0.0001	0.0021
34	0.80	-0.0011	-0.0000	0.0033	0.0000	0.0033
35	0.81	0.0011	-0.0001	0.0032	0.0001	0.0031
36	0.90	0.0011	-0.0006	0.0006	0.0001	0.0002
37	1.00	0.0011	-0.0010	-0.0028	-0.0001	-0.0039
38	1.10	0.0011	-0.0013	-0.0027	-0.0000	-0.0040
39	1.20	0.0011	-0.0014	0.0009	0.0001	-0.0004
40	1.30	0.0011	-0.0013	0.0033	-0.0001	0.0019
41	1.40	0.0011	-0.0009	0.0015	-0.0001	0.0005
42	1.50	0.0011	-0.0005	-0.0022	0.0001	-0.0025
43	1.57	0.0011	-0.0001	-0.0033	0.0000	-0.0034
44	1.58	0.0000	-0.0000	-0.0033	-0.0000	-0.0033

FIGURE 2.22. Fourier series.

```
C2:   'FOURIER SERIES
A16:  'Phase Angle,rads
C16:  @ATAN(C14*$C$6*$C$10/($C$9-(C14^2*$C$6^2*$C$8)))
D16:  @ATAN(D14*$C$6*$C$10/($C$9-(D14^2*$C$6^2*$C$8)))
E16:  3.14159-@ATAN(E14*$C$6*$C$10/($C$9-
          (E14^2*$C$6^2*$C$8)))
A17:  'Phase Angle,deg
C17:  @DEGREES(C16)
D17:  @DEGREES(D16)
E17:  @DEGREES(E16)
A18:  'Amplitude,m
C18:  4*$C$9*$C$5/(3.4159*C14*(($C$9-(C14^2*$C$6^2*$C$8))^2
          +(C14*$C$6*$C$10)^2)^0.5)
D18:  4*$C$9*$C$5/(3.4159*D14*(($C$9-(D14^2*$C$6^2*$C$8))^2
          +(D14*$C$6*$C$10)^2)^0.5)
E18:  4*$C$9*$C$5/(3.4159*E14*(($C$9-(E14^2*$C$6^2*$C$8))^2
          +(E14*$C$6*$C$10)^2)^0.5)
C20:  'Calculation of Harmonics
A25:  (F2) 0
B25:  -$C$5
C25:  +$C$18*@SIN($C$14*$C$6*A25-$C$16)
D25:  +$D$18*@SIN($D$14*$C$6*A25-$D$16)
E25:  +$E$18*@SIN($E$14*$C$6*A25-$E$16)
F25:  @SUM(C25..E25)
A26:  (F2) 0.1
B26:  -$C$5
C26:  +$C$18*@SIN($C$14*$C$6*A26-$C$16)
D26:  +$D$18*@SIN($D$14*$C$6*A26-$D$16)
E26:  +$E$18*@SIN($E$14*$C$6*A26-$E$16)
F26:  @SUM(C26..E26)
A41:  (F2) 1.6
B41:  +$C$5
C41:  +$C$18*@SIN($C$14*$C$6*A41-$C$16)
D41:  +$D$18*@SIN($D$14*$C$6*A41-$D$16)
E41:  +$E$18*@SIN($E$14*$C$6*A41-$E$16)
F41:  @SUM(C41..E41)
```

FIGURE 2.23. Fourier cell formulas.

represent the estimate of vibration problem caused by a triplex pump, what is the maximum amplitude and frequency of the worst case harmonic? The Periodic function is assumed to be a square wave. The square wave is also assumed to be symmetrical about the origin and would use the Fourier Series for an odd function. The example will be worked using the more cumbersome trigonometric series rather than the complex series to illustrate the trignometric capability of Quattro Pro. The expansion of the Fourier Series will be limited to the 5th harmonic since the higher harmonics are usually of small magnitude.

The motion of the upper end of the spring approximates a square wave with a fundamental frequency of 0.636 Hz. The time of the cycle is 1.5723 seconds and the amplitude is 0.0011 meters. The square wave is a function

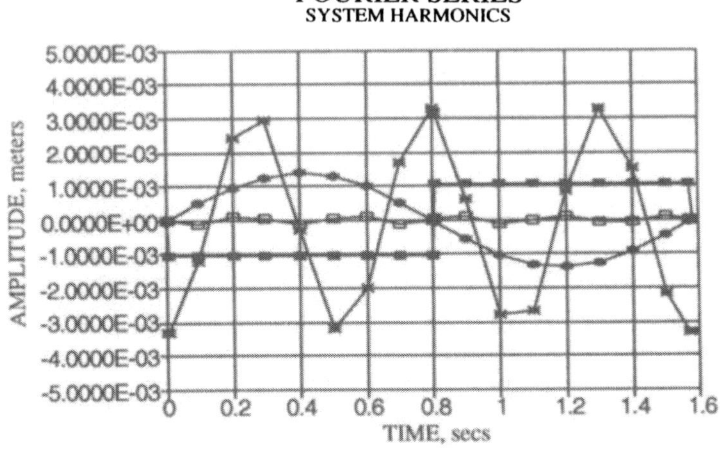

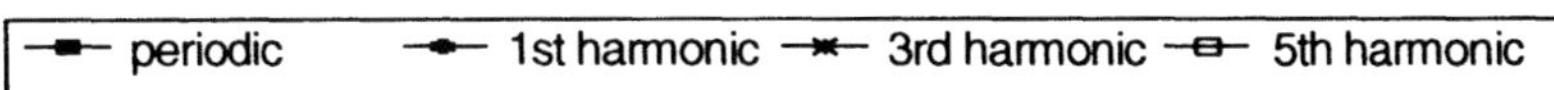

FIGURE 2.24. System harmonics.

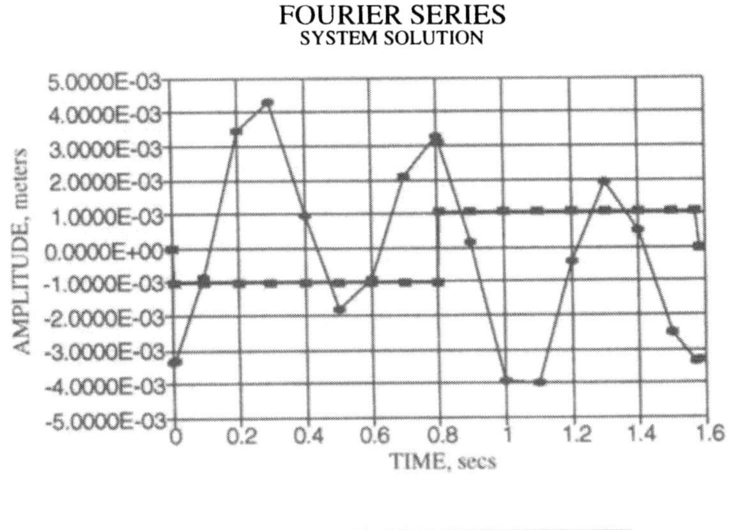

FIGURE 2.25. System movement.

of time by the sum of the series:

$$x = \frac{4Y}{\pi} \sum_{n=1,3,5..}^{\infty} \frac{\sin nwt}{n}$$

$$n = \text{odd harmonics}, 1, 3, 5 \cdots \infty. \tag{1}$$

The phase angle a_n and the amplitude y_n of the nth harmonic at any instant is written for the system as:

$$a_n = \tan^{-1} \frac{nwS}{K - n^2 w^2 M} \tag{2}$$

$$y_n = \frac{4KY}{\pi((k - n^2 wM)^2 + (nwS)^2)^{0.5}}. \tag{3}$$

From the above equations, the phase angle and the amplitude of each harmonic can be calculated and substituted into:

$$x_n = \sum_{n=1,3,5..}^{\infty} y_n \sin(nwt - a_n). \tag{4}$$

Each of the harmonic waves can be calculated and plotted from the above equation. The motion of the mass is the algebraic sum of each of the harmonics.

The spreadsheet in Figure 2.22 is laid out to show the calculation of the phase angle and the amplitude of the harmonics. The remainder of the spreadsheet is more conventional in the way it calculates the harmonic wave shapes. The mass motion is shown in column F.

The cell formulas in Figure 2.23 have been trimmed to show only the basic formulas. It should be pointed out that the use of the relative and absolute cell addresses, as well as the use of the copy command, makes the generation of the equations quite easy.

The graphic plot of the system harmonics in Figure 2.24 shows the effect of the phase angle and amplitude on the harmonics. The mass movement is plotted as the algebraic sum of the harmonic displacements in Figure 2.25.

3
Matrix Algebra

Matrix operations are a problem solving tool in all branches of science and engineering. Matrices can be used in the calculation of primary stresses of a truss. The analysis of electronic circuits relies heavily on the use of matrices. Matrix algebra was used by Heisenberg in a parallel development of quantum mechanics: in fact the Schrödinger wave equation and the matrix mechanics of Heisenberg are mathematically equivalent.

The Advanced Math menu has two commands, Invert and Multiply, available to perform matrix algebra. There are limitations to the matrix commands: matrix addition, matrix substraction and scalar multiplication are not performed, and inversion is limited to square matrix. In a square matrix, the number of rows equals the number of columns. The multiplication is performed only if the number of columns of the first matrix and the number of rows of the second matrix are equal.

Quattro Pro will send an error message if the matrix is singular. A singular matrix contains a redundancy or inconsistency. A redundancy occurs when one of the equations is a variation of another equation. An inconsistency is when two equations may appear equal but give conflicting results.

Matrices are used in the solution of linear simultaneous equations. If the constraints are defined as the equations and the variables as the unknowns, then for Quattro Pro to solve the matrix, the number of variables must equal the number of constraints. Each row of the matrix corresponds to a linear constraint and each column to one variable.

If a linear system is defined as:

$$x - y + 5z = -6$$
$$3x + 3y - z = 10$$
$$x + 3y - 2z = 5$$

The variable coefficients are set up as matrix A, the constants as matrix B and the variables as matrix X. Solving for matrix X would involve:

$$AX = B$$

	A	B	C	D	E
1					
2		SAMPLE PROBLEM			
3		MATRIX ALGEBRA			
4					
5	--				
6		A		B	X
7	--				
8	1	-1	5	-6	x
9	3	3	-1	10	y
10	1	3	2	5	z
11					
12	--				
13		1/A		X	
14	--				
15	0.195652	0.369565	-0.30435	x =	1
16	-0.15217	-0.06522	0.347826	y =	2
17	0.130435	-0.08696	0.130435	z =	-1

FIGURE 3.1. Sample problem spreadsheet.

Since direct division is not possible in matrix algebra, division is accomplished by inversion of matrix A and multiplying. The linear system is set up as follows:

$$A = \begin{vmatrix} 1 & -1 & 5 \\ 3 & 3 & -1 \\ 1 & 3 & 2 \end{vmatrix} \quad B = \begin{vmatrix} -6 \\ 10 \\ 5 \end{vmatrix} \quad X = \begin{vmatrix} x \\ y \\ z \end{vmatrix}$$

The coefficients in A are a 3×3 matrix. Both B and X are designated as 3×1 matrices. The matrices are set up on the spreadsheet with a value in each cell as in Figure 3.1. The inverted matrix 1/A is sent to an output block which is specified by the top left cell. The multiplication of the matrices (1/A)*B is sent to a variables block specified by the top cell. If there are data in the output block cells, it will be overwritten.

Truss Design

A simple truss design has been chosen to illustrate the use of matrices to calculate the primary stresses. The truss is made up of members lying in a plane. The members of bars are joined only at the ends. Frictionless pins through holes on the axes are assumed at the junction of the bars. The loads, which lie in the plane of the truss, are applied to the pins. By this arrangement only direct compression or tension exists in any member.

The actual stresses due to the size and weight of the truss, bending, and rigidity of the joints are called the secondary stresses. The major secondary stresses are not included in this example.

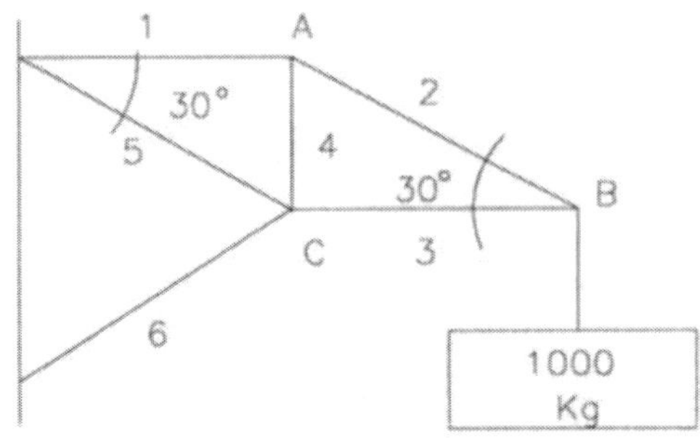

FIGURE 3.2. Drawing of truss.

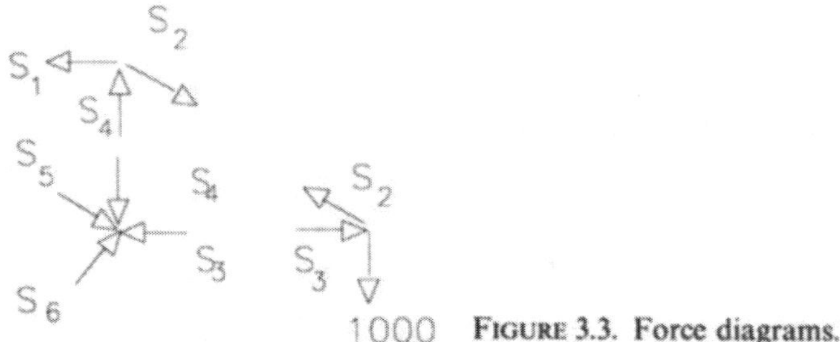

FIGURE 3.3. Force diagrams.

It should be noted that in the truss in Figure 3.3, the angles only of the members are used in the determination of the primary stresses. The compression and tension stresses are then used to size the members.

In the calculation of the primary stresses, the algebraic sum of the stresses in the x direction and in the y direction must equal zero. Each joint contributes two equations to the simultaneous system. By convention, the tension stresses are positive and the compression stresses negative.

The x and y force equations (Figure 3.3) are shown below:

$$x \text{ forces, point 1: } -\cos(30)S_6 + S_3 - \cos(30)S_5 = 0$$
$$y \text{ forces, point 1} \quad \sin(30)S_6 - \sin(30)S_5 - S_4 = 0$$

$$x \text{ forces, point 2} \quad -S_3 - \cos(30)S_2 = 0$$
$$y \text{ forces, point 2} \quad -\sin(30)S_2 + 1000 = 0$$

$$x \text{ forces, point 3} \quad \cos(30)S_2 - S_1 = 0$$
$$y \text{ forces, point 3} \quad S_4 + \sin(30)S_2 = 0.$$

The force equations then are rearranged as simultaneous equations:

$$+\sin(30)S_6 \quad +0 \quad -\sin(30)S_5 \quad -S_4 \quad +0 \quad +0 = 0$$
$$+0 \quad -S_3 \quad +0 \quad +0 - \cos(30)S_2 \quad +0 = 0$$
$$-\cos(30)S_6 \quad +S_3 - \cos(30)S_5 \quad +0 \quad +0 \quad +0 = 0$$
$$+0 \quad +0 \quad +0 \quad +1 + \sin(30)S_2 \quad +0 = 0$$
$$+0 \quad +0 \quad +0 \quad +0 - \sin(30)S_2 \quad +0 = -1000$$
$$+0 \quad +0 \quad +0 \quad +0 + \cos(30)S_2 \quad -S_1 = 0.$$

	A	B	C	D	E	F	G	H
1								
2			**TRUSS DESIGN**					
3	-------							-------
4				**A**			**X**	**B**
5	-------							-------
6	0.5	0	-0.5	-1	0	0	S_6	0
7	0	-1	0	0	-0.866	0	S_3	0
8	-0.866	1	-0.866	0	0	0	S_5	0
9	0	0	0	1	0.5	0	S_4	0
10	0	0	0	0	-0.5	0	S_2	-1000
11	0	0	0	0	0.866	-1	S_1	0
12								
13								
14	-------							-------
15				**1/A**			**X**	**= Kg**
16	-------							-------
17	1	-0.577	-0.577	1	2	0	$S_6=$	-2000
18	1	0	0	0	1.732	0	$S_3=$	-1732
19	-1	-0.577	-0.577	-1	0	0	$S_5=$	0
20	0	0	0	1	1	0	$S_4=$	-1000
21	0	0	0	0	-2	0	$S_2=$	2000
22	0	0	0	0	-1.732	-1	$S_1=$	1000

FIGURE 3.4. Truss design spreadsheet.

The simultaneous equations can be set up as a matrix problem to be solved for $AX = B$ as in Figure 3.4.

The trigonometric formulas used are in degrees. Since Quattro Pro only recognizes radians, the functions are written:

$$SIN\,30° = @\,SIN(@\,RADIANS(30))$$
$$COS\,30° = @\,COS(@\,RADIANS(30)).$$

The C member is not stressed and it is not needed. The calculation of the primary or tension and compression stresses often indicates the presence of an extra member or possibly the need for an additional member.

DC Network

In electrical engineering a combination of electrical components that provide a number of parrallel and series circuits is called a network. The direct current resistive network will involve only the use of real numbers. If the network has alternating current or has capacitance and/or inductance, the solution usually will require complex numbers.

Quattro Pro is limited to real numbers in the matrix algebra program. Complex numbers such as $x + jy$, where x is real and y is imaginary, are not within the matrix algebra capabilities.

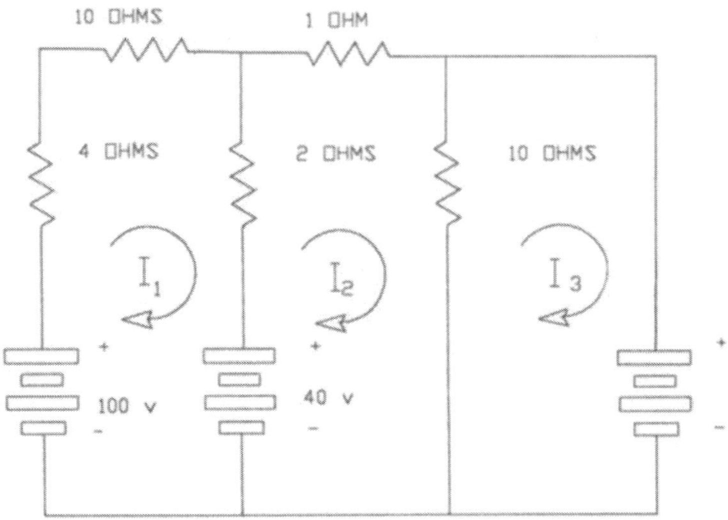

FIGURE 3.5. Resistive network.

	A	B	C	D	E
1					
2		Matrix Arithmetic			
3					
4	----	----	----	----	----
5		A		B	X
6	----	----	----	----	----
7	16	-2	0	60	I1
8	-2	13	-10	40	I2
9	0	-10	10	-20	I3
10					
11	----	----	----	----	----
12	1/A				X
13	----	----	----	----	----
14	0.068182	0.045455	0.045455	I1 =	5
15	0.045455	0.363636	0.363636	I2 =	10
16	0.045455	0.363636	0.463636	I3 =	8

FIGURE 3.6. Network spreadsheet.

Figure 3.5 is a direct current resistive network with only three circuits. It is adequate to illustrate the use of matrix arithmetic for the solution of a system of linear equations but still can be solved by inspection.

Kirchoff's voltage law states that the algebraic sum of the voltages around the loop of a circuit is zero. The voltages around loops I_1, I_2, and I_3 are:

$$100 - 4I_1 - 10I_1 - 2(I_1 - I_2) - 40 = 0$$
$$40 - 2(I_2 - I_1) - I_2 - 10(I_2 - I_3) = 0$$
$$-20 - 10I_2 + 10I_3 = 0$$

These equations simplify to and become the network system linear simultaneous equations:

$$16I_1 - 2I_2 + 0 = 60$$
$$-2I_1 + 13I_2 - 10I_3 = 40$$
$$0 - 10I_2 + 10I_3 = -20$$

The matrices then can be written as:

$$A = \begin{vmatrix} 16 & -2 & 0 \\ -2 & 10 & -10 \\ 0 & -10 & 10 \end{vmatrix} \quad B = \begin{vmatrix} 60 \\ 40 \\ -20 \end{vmatrix} \quad X = \begin{vmatrix} I_1 \\ I_2 \\ I_3 \end{vmatrix}$$

Matrix A is placed in block A7···C9. Using the command of /**Tools**| **Advanced Math**|**Invert** for Matrix A, the inverted matrix 1/A is located in block A14···C16. Multiplication of 1/A*B is done with /**Tools**|**Advanced Math**|**Multiply** and placed in the block E14···E16

From the spreadsheet in Figure 3.6, the currents in the loops are:

$$I_1 = 5, \quad I_2 = 10, \quad \text{and} \quad I_3 = 8 \text{ amperes.}$$

4
Curve Fitting with Linear Regression

With test data it is often necessary to determine a relationship between the two sets of numbers, for interpolation and extrapolation. It is possible to graph the data as linear, semi-log or log plots to determine the equation and constants. The XY line then is a visual estimate of the average point positions. A more accurate method is to determine the equation of the graph of data statistically by using linear regression. The linear regression method can be used not only for linear data but for any type of curve that can be rewritten in a linear form. The tabulation below is for the most common equations:

Curve	Equation	Linearized equation
Linear	$y = a + bx$	
Exponential	$y = ae^{bx}$	$\ln y = \ln a + bx$
Natural log	$y = a + b \ln x$	
Power	$y = ax^b$	$\ln y = \ln a + b \ln x$

Most calculator and computer curve fittings programs are written to be entered with the data as recorded. The data are manipulated to fit the needs of the linearized equation and a linear regression analysis is performed for each curve. Some programs will automatically select the curve with the R-squared value closest to 1.000. Quattro Pro utilizes only the linear regression program, so to analyze the complex curves the data must be submitted in the form of the linearized equation such as the natural log. The output data will then be in the same form.

Under Advanced Math of the Main menu is the Regression command. With this built in command, Quattro will provide data correlation. Each of the above curves will be illustrated by calculating the y data from the equation, running a linear regression, calculating the y values from the Quattro Pro constants and making a residual comparison. After demonstrating the accuracy of the regression algorithm, an equation will be calculated from data for use in a later example.

60

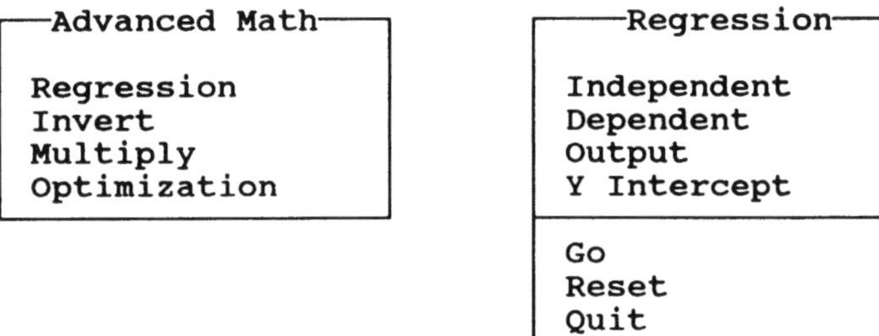

FIGURE 4.1. Advanced math and regression menus.

From the Advanced menu and the Regression Menu, Figure 4.1, the column that is to be the independent variable must be selected. The column that is to be the dependent variable is specified. These columns must be the same length, i.e., ordered pairs. The position of the output block is selected as to not cverwrite data. The Y intercept will force the y value to 0.

On the following pages, Figures 4.2, 4.5, 4.8, and 4.11 describe the regression analyses. To facilitate the location of an individual cell formula, the spreadsheet formulas are shown in Figures 4.3, 4.6, 4.9 and 4.12. The graphs of the curves are Figures 4.4, 4.7, 4.10 and 4.13.

```
        A           B           C               D
 1
 2              LINEAR  CURVE  FIT
 3               y  =  a  +  bx
 4               y  =  3  +  2x
 5----------------------------------------------
 6      x            y        calc y        error
 7----------------------------------------------
 8      2            7           7            0
 9      4           11          11            0
10      6           15          15            0
11      8           19          19            0
12     10           23          23            0
13     12           27          27            0
14
15                     a  =  3
16                     b  =  2
               Regression  Output
        Constant                              3
        Std Err of Y Est              3.49E-07
        R Squared                             1
        No. of Observations                   6
        Degrees of Freedom                    4
        X Coefficient(s)          2
        Std Err of Coef.    4.17E-08
```

FIGURE 4.2. Linear spreadsheet.

The same constants are used for each of the equations in the example:

$$a = 3$$
$$b = 2$$

Straight Line. Since the equation is in the linear form, it is used directly, with x as the independent and y as the dependent variable. The Output shows the constant (which is a) as 3 and the x coefficient (which is b) as 2. The R squared is 1.0 indicating the best possible fit. The residual is the difference between y and the calculated y and is 0 for all cases.

Exponential Curve. The equation is $y = ae^{bx}$ and the linear form is $\ln y = \ln a + bx$. The independent variable is the direct value of x. The dependent variable, y, is input as the $\ln y$ as in the linear equation. The output will show the constant as the ln of a. The X coefficient will be the numerical value of b. The R squared again is 1.0 and the residual is calculated as 0. While the graph in Figure 4.7 is an obvious distortion due to the selection of the variables, the Quattro Pro linear regression provides an accurate track of the original equation.

Natural Log Curve. The equation is $y = a + b \ln x$, which is already in linear form. The independent variable x is input as the ln x. The output gives the constants as the numerical values of a and b. R squared again is 1.0 and the residual is 0.

Power Curve. The equation is $y = ax^b$ and the linear form is $\ln y = \ln a + b \ln x$. The variables x, y and a are input as the natural log and the output gives a as the log and b as the numerical value. Again R squared is 1.0 and the residual is 0.

```
B2:  'LINEAR CURVE FIT
B8:  3+(2*A8)
C8:  3+(2*A8)
B9:  3+(2*A9)
C9:  3+(2*A9)
B10: 3+(2*A10)
C10: 3+(2*A10)
B11: 3+(2*A11)
C11: 3+(2*A11)
B12: 3+(2*A12)
C12: 3+(2*A12)
B13: 3+(2*A13)
C13: 3+(2*A13)
```

FIGURE 4.3. Linear spreadsheet formulas.

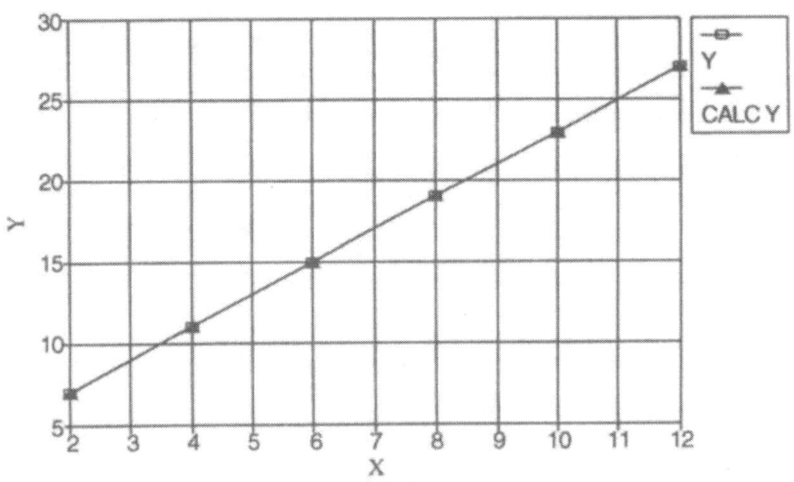

FIGURE 4.4. Linear graph.

	A	B	C	D	E
1					
2		EXPONENTIAL CURVE FIT			
3		y = a*exp bx			
4		y = 3*exp(2x)			
5		--			
6	x	y	ln y	calc y	error
7		--			
8	2	163.7945	5.098612	163.7945	0
9	4	8942.874	9.098612	8942.874	0
10	6	488264.4	13.09861	488264.4	0
11	8	26658332	17.09861	26658332	0
12	10	1.46E+09	21.09861	1.46E+09	0
13	12	7.95E+10	25.09861	7.95E+10	0
14					
15		a = exp (1.098612) = 3			
16		b = 2			

```
                 Regression Output
         Constant                      1.098612
         Std Err of Y Est              1.52E-07
         R Squared                            1
         No. of Observations                  6
         Degrees of Freedom                   4
         X Coefficient(s)                     2
         Std Err of Coef.              1.82E-08
```

FIGURE 4.5. Exponential spreadsheet.

```
 B2:  'EXPONENTIAL CURVE FIT
 B8:  3*(@EXP(2*A8))
 C8:  @LN(B8)
 D8:  3*@EXP(2*A8)
 B9:  3*(@EXP(2*A9))
 C9:  @LN(B9)
 D9:  3*@EXP(2*A9)
B10:  3*(@EXP(2*A10))
C10:  @LN(B10)
D10:  3*@EXP(2*A10)
B11:  3*(@EXP(2*A11))
C11:  @LN(B11)
D11:  3*@EXP(2*A11)
B12:  3*(@EXP(2*A12))
C12:  @LN(B12)
D12:  3*@EXP(2*A12)
B13:  3*(@EXP(2*A13))
C13:  @LN(B13)
D13:  3*@EXP(2*A13)
E15:  @EXP(E18)
```

FIGURE 4.6. Exponential spreadsheet formulas.

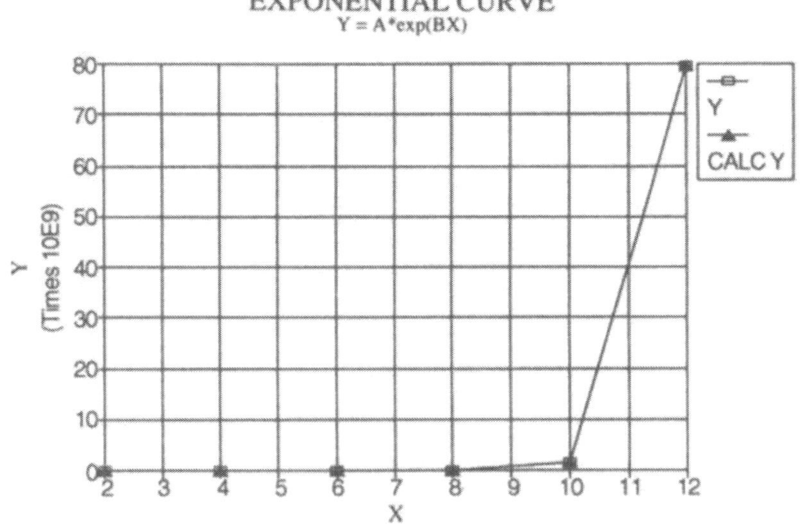

FIGURE 4.7. Exponential graph.

```
        A         B          C          D          E
 1
 2              NATURAL-LOG  CURVE
 3                  y = a + b lnx
 4                  y = 3 + 2*lnx
 5    ------------------------------------------------
 6       x          y         lnx      calc y  residual
 7    ------------------------------------------------
 8       2  4.386294  0.693147  4.386294         0
 9       4  5.772589  1.386294  5.772589         0
10       6  6.583519  1.791759  6.583519         0
11       8  7.158883  2.079442  7.158883         0
12      10  7.60517   2.302585   7.60517         0
13      12  7.969813  2.484907  7.969813         0
14
15                    a =      3
16                    b =      2
17                    Regression Output:
             Constant                        3
             Std Err of Y Est         1.83E-07
             R Squared                       1
             No. of Observations             6
             Degrees of Freedom              4

             X Coefficient(s)                2
             Std Err of Coef.         1.24E-07
```

FIGURE 4.8. Natural log spreadsheet.

```
B2:  'NATURAL-LOG  CURVE
B8:  3+2*@LN(A8)
C8:  @LN(A8)
D8:  3+2*@LN(A8)
B9:  3+2*@LN(A9)
C9:  @LN(A9)
D9:  3+2*@LN(A9)
B10: 3+2*@LN(A10)
C10: @LN(A10)
D10: 3+2*@LN(A10)
B11: 3+2*@LN(A11)
C11: @LN(A11)
D11: 3+2*@LN(A11)
B12: 3+2*@LN(A12)
C12: @LN(A12)
D12: 3+2*@LN(A12)
B13: 3+2*@LN(A13)
C13: @LN(A13)
D13: 3+2*@LN(A13)
```

FIGURE 4.9. Natural log spreadsheet formulas.

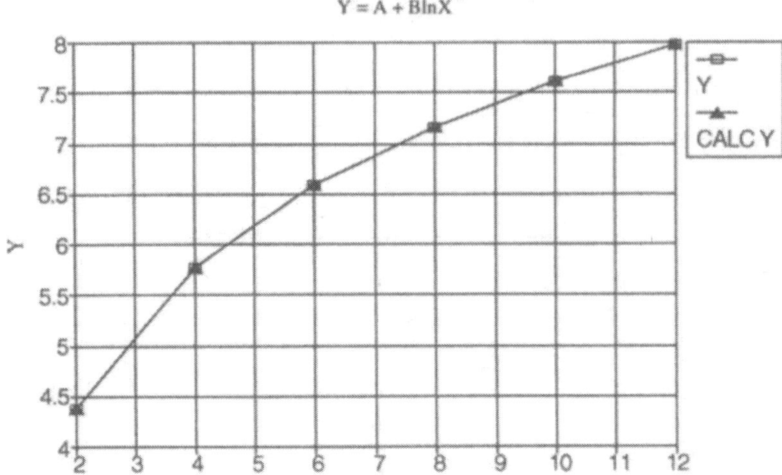

FIGURE 4.10. Natural Log graph.

	A	B	C	D	E	F
1						
2			POWER CURVE FIT			
3		y = a x ^b	ln y = ln a + b ln x			
4		ln y = ln 3 + 2*ln x				
5	----	-----	----------	----------	--------	----------
6	x	y	ln x	ln y	calc y	residual
7	----	-----	----------	----------	--------	----------
8	2	12	0.693147	2.484907	12	0
9	4	48	1.386294	3.871201	48	0
10	6	108	1.791759	4.682131	108	0
11	8	192	2.079442	5.257495	192	0
12	10	300	2.302585	5.703782	300	0
13	12	432	2.484907	6.068426	432	0
15						
16			a = 3			
17			b = 2			

Regression Output:

Constant	1.098612
Std Err of Y Est	9.06E-08
R Squared	1
No. of Observations	6
Degrees of Freedom	4
X Coefficient(s)	2
Std Err of Coef.	6.11E-08

FIGURE 4.11. Power spreadsheet.

```
                        B2:  'POWER CURVE FIT
     B8:  3*(A8^2)                          B12:  3*(A12^2)
     C8:  @LN(A8)                           C12:  @LN(A12)
     D8:  @LN(B8)                           D12:  @LN(B12)
     E8:  3*A8^2                            E12:  3*(A12^2)
     B9:  3*(A9^2)                          B13:  3*(A13^2)
     C9:  @LN(A9)                           C13:  @LN(A13)
     D9:  @LN(B9)                           D13:  @LN(B13)
     E9:  3*A9^2                            E13:  3*(A13^2)
    B10:  3*(A10^2)
    C10:  @LN(A10)
    D10:  @LN(B10)
    E10:  3*A10^2
    B11:  3*(A11^2)
    C11:  @LN(A11))
    D11:  @LN(B11))
    E11:  3*(A11^2)
```

FIGURE 4.12. Power spreadsheet formulas.

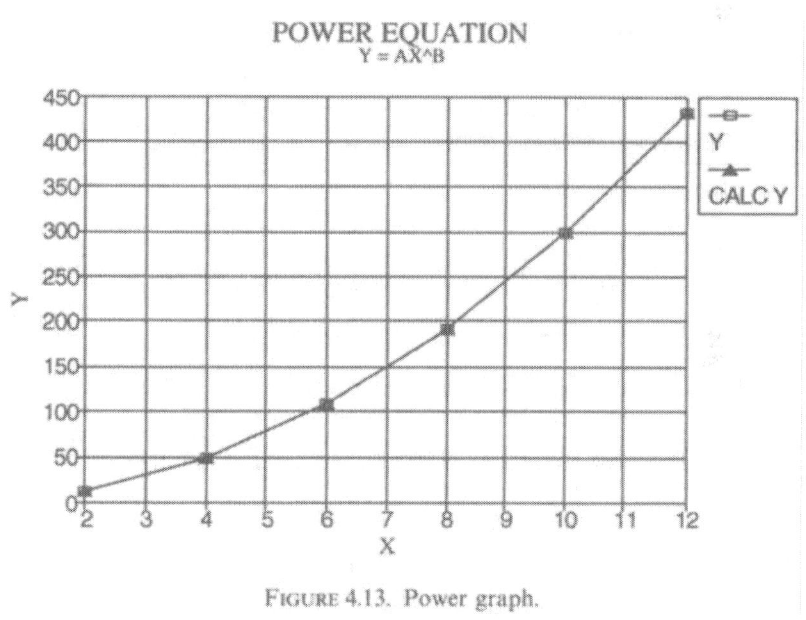

FIGURE 4.13. Power graph.

Damping Force Data Curve

As a more practical example, the following damping force data for the numerical solution of a differential equation will be evaluated to determine the type of curve and the constants for the function. The equation is to be

```
            EXPONENTIAL CURVE FIT
               Y = A*EXP(BX)
              ln Y = ln A + BX
    ---------------------------------------------
       X          Y        ln Y      CALC Y   RESIDUAL
    ---------------------------------------------
      0.0        1.0                 0.90821   0.09179
      0.6        1.1      0.09531    1.15952  -0.05952
      1.2        1.4      0.33647    1.48038  -0.08037
      1.9        1.9      0.64185    1.96855  -0.06855
      2.5        2.5      0.91629    2.51327  -0.01327
      3.1        3.3      1.19392    3.20872   0.09127
      3.7        4.2      1.43509    4.09661   0.10338

      A = @ exp(-0.09628)=   0.90821
      B =                    0.40715
              Regression Output:
    Constant                     -0.09628
    Std Err of Y Est              0.059758
    R Squared                     0.990141
    No. of Observations                  7
    Degrees of Freedom                   5

    X Coefficient(s)              0.407146
    Std Err of Coef.              0.018169
```

FIGURE 4.14. Exponential curve fit spreadsheet.

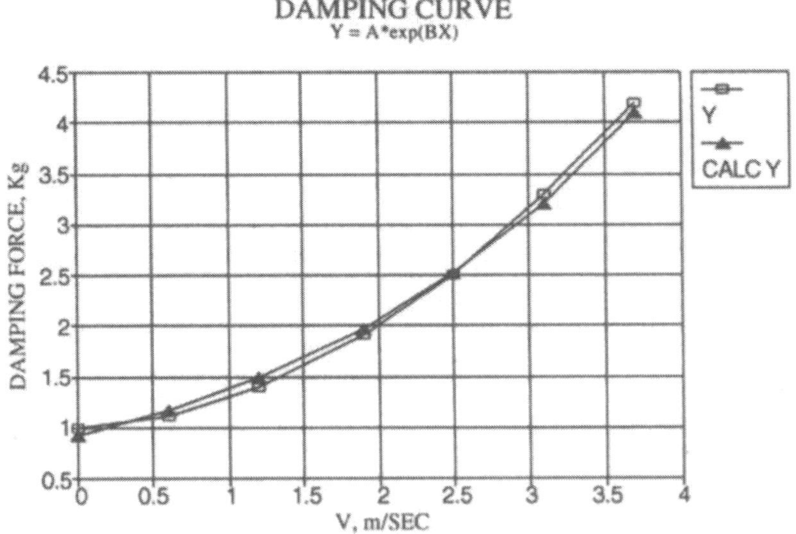

FIGURE 4.15. Damping curve graph.

used in a later example.

v, m/sec	3.7	3.1	2.5	1.9	1.2	0.6	0.0
damping force, kg	4.2	3.3	2.5	1.9	1.4	1.1	1.0

Three of the equations use the log of one or more variables which places limits on x and y. Any variable that is shown as a log in the linearized equation must not be 0 or a negative value. Since the data are obviously not linear, and the data contains a 0, all but the exponential curve is eliminated.

From the above spreadsheet, $\ln A = -0.09628$ or $A = 0.9082$ and $B = 0.4072$ with R squared as 0.9901. The final equation is:

$$y = 0.9082e^{0.4072x}$$

5
Database

In addition to the spreadsheet integrated software, Quattro Pro has the added dimension of a database. The dedicated database programs that relate the data in several different files based upon comman elements are usually a programmable relational database. Quattro Pro is not a relational database. The entire database is held in the memory by Quattro Pro. The larger the Quattro Pro database, the larger the RAM requirement, but the access and processing time is reduced.

As a review, a record is defined as a set of related data. In the database, the record is the row. The fields are the columns, each dedicated to similar data. For example, a set of test data would be the fields, such as temperature, pressure, and composition. The rows or records would be for each set of fields, indicating time, number of test, or a similar parameter.

The format of the database is more restrictive than that of the spreadsheet. The field names (column headings) cannot be separated from the first data row (record) by a blank or decorative line. The field name can consist of only a single word of no more than 15 characters. If a spreadsheet is set up using the database format, the spreadsheet data can also be manipulated as a database.

Quattro Pro can store up to 8191 records with 256 fields containing more than 250 characters in each field. Quattro Pro is smaller than a dedicated database, but is quite adequate to store and manipulate data from the average experiment or test. If the test required that ten data items or fields be listed for each step or record, then over 800 steps or records can be made, provided that the memory is available.

As part of the spreadsheet, the database commands are similar to those of the spreadsheet. The entering and modification of the records are the same as the spreadsheet.

Sort

A database can only sort or change the rows or records. The records can be arranged in a specific order defined by one to five different fields. The records are sorted first by the primary sort field, then by the secondary sort

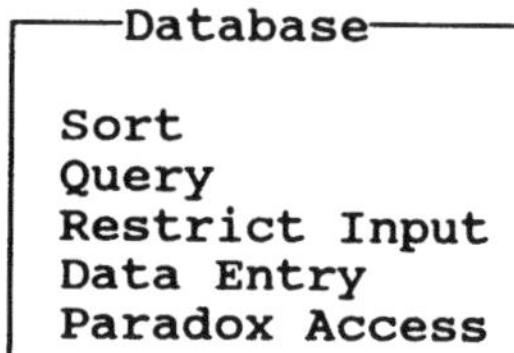

FIGURE 5.1. Database menu.

FIGURE 5.2. Sort menu.

field and so sequentially up to the fifth field. For example, if the primary sort is temperature, and more than one of the same temperature is encountered, then each set of duplicate temperatures will be sorted in sequence determined by another parameter of the secondary sort field.

/**Database**|**Sort** will bring up the Sort menu of Figure 5.2. The first step is to select Block and identify the cell block to be sorted. It is not necessary for all records or rows to be in the cell block. However, all the cells of each record to be sorted must be included, or the fields of the database will be mismatched. The cell block should not include the field names or column headings.

Sort Keys

The field to provide the primary sorting data is selected by placing the cursor on a cell in the column. When the 1st Key is pressed, Quattro Pro requests for either a descending or an ascending order. Command is returned to the Sort menu. If further sorting is required, the second to the fifth keys can be assigned to the fields. Go must be pressed to start the sort. The database is reordered and the Sort menu is returned.

The Undo (Alt-F5) can reverse an erroneous sort, provided that the Undo command was enabled before it was needed. The addition of a column of sequential numbers before sorting can be used to resort the database to the original order. The numbers can be added by inserting a blank column on the left and using the /**Edit**|**Fill** command.

Reset

If it is decided that an additional or successive sort must be made, Quattro Pro will remember the sort block and the sort keys. This is convenient if the new sort is to be an extension of the previous sort. The block is still defined and an unused key is assigned to the new field. If, however, a previously used key must be reassigned, the Reset option must be used. Reset not only clears the keys but also unlocks the block coordinates as well.

Sort Rules

There two sort directions, ascending and descending. The default sort conditions are in the following order:
Ascending Order:

1. blank cells
2. in numerical order for labels beginning will numbers
3. in ASCII order for labels beginning with letters and special characters
4. values are sorted in numerical order

Descending Order:

1. values are sorted in reverse numerical order
2. in reverse ASCII order for labels beginning with letters or special characters
3. in reverse numerical order for labels beginning with numbers
4. blank cells

The command /**Database**|**Sort**|**Sort Rules** brings up the Sort Rules menu of Figure 5.3. The selection of the Numbers Before Labels will set Quattro Pro to sort the numbers before or after the labels.

The Label Order provides a menu with the choice of ASCII or Dictionary sort. The ASCII sort follows the ASCII code of the first letter. The Dictionary sort disregards case and sorts alphabetically.

Search

The Quattro Pro database provides the ability to quickly search the database for specific data records. With a minimum of keystrokes the records based upon a set of criteria can be found and copied to another area of the

```
┌─────────Sort Rules─────────┐
│                            │
│   Numbers Before Labels    │
│   Label Order              │
│   Quit                     │
└────────────────────────────┘
```

FIGURE 5.3. Sort rules menu.

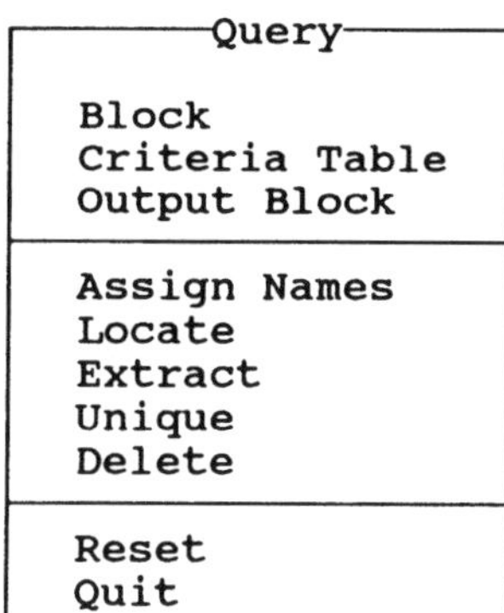

FIGURE 5.4. Query menu.

spreadsheet for review of printing. Specific records can be deleted or duplicate records located and all but one deleted. To implement a search the /**Database**|**Query** menu is called up.

Block

The block of data that is to be searched can be the entire database or it can be any part. No matter what the size of the block, it must include the field names of the columns to be searched. /**Database**|**Query**|**Block** will bring up the Query menu and the Block entry. The block can be defined either by entering the coordinates, or by moving the cell selector. For example, if the cell selector were moved to the upper left cell and anchored by pressing a period, moving the cell selector to the lower right of the block and anchoring the cell will set the block. ENTER is then pressed to confirm the block selection.

If the database block is another open file, it can be moved to the present database with the linking syntax. Entering Block will cause Quattro Pro to ask for the cell positions of the block. At this point, the following is an example of the link request.

[FILE NAME]A1 ··· D20

Criteria Table

The criteria table should be located in an unused portion of the database. After using /**Database**|**Query**|**Criteria Table** the field names of the search block are copied into the first row. Below the field name enter the data to be found. If a condition is required, the search formula under the field name must include the cell reference, an operator and a value. If the Assigned Names has not been actuated, the cell reference must be the first entry in the field preceded by a plus (+).

Any of the logic operators can be used in the search formulas. Two or more search formulas can be tied together in one field by using #AND#, #OR# and #NOT#. Wildcards may be used in the criteria table. The question mark (?) replaces a single character. The asterisk (*) is a multiple character wildcard. The tilde (~) is used to list those labels that do not match a search condition.

The entries in a criteria table that are on the same line are related by AND. The record must meet both criteria to be included in the Locate operation. If the entries are on different lines, the relation is OR. Any one of the criteria will then provide a match for the Locate command.

Output Block

If any of the commands that will print the search data are to be used, an output block must be designated. The area must not overlap the spreadsheet or the criteria table. The first row of the block must contain the names of the fields that are in the output of the database.

Assign Names

Using /**Database**|**Query**|**Assign Names**, Block names can be assigned to each cell in the second row by using the field name of the first row. A table of the block names and the cell location can be made in an unused portion of the spreadsheet. To create the table, after assigning the names, use the /**Edit**|**Names**|**Make Table** command.

Locate

After setting up the Block, Output Block and the Criteria, the /**Database**|**Query**|**Locate** command will highlight the first record in the database that meets the criteria. To edit the record use the arrows to move the cell locator until the field to be changed is highlighted. Pressing F2 will enter the Edit mode. The changes are entered by pressing ENTER.

Extract

The /**Database**|**Query**|**Extract** command will locate and copy all records that meet the criteria to the Output Block. Only the field entries that are listed in the Output Block will be displayed. If the output block is too small, a warning is displayed and pressing ESC will return to the Query menu. The size of the Output Block then can be changed.

Unique

The /**Database**|**Query**|**Unique** command will not display duplicates of selected records. A duplicate must be identical in all fields, not just those specified by the criteria table.

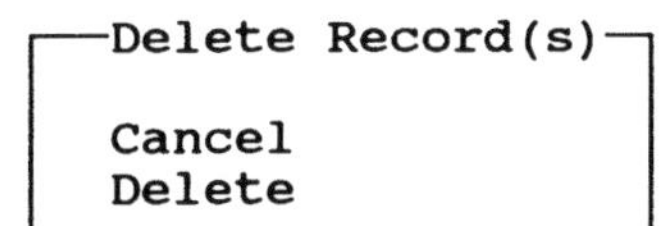

FIGURE 5.5. Delete records menu.

Delete

Figure 5.5 is the /**Database**|**Query**|**Delete** menu. In the Delete menu, confirmation is requested before completing the command.

The operation can be Cancelled by pressing C. The deletion is made by pressing D.

Restrict Input

The Restrict Input commands are most useful for a database macro that is set up for technician data input. The access is limited to only those cells in a block that are not protected.

Data Entry

The Data Entry command can be used in a database macro to require cells only to accept labels or dates and is of use in limiting errors.

Database

The influent and effluent oil and grease analyses of a sewage treatment plant are stored in a database, Figure 5.6. A daily analysis for the total oil and grease is required of an effluent sample taken with a 24 hour continuous composite sampler. The analysis is to be made using an approved method with an assigned lower detection limit of 5.0 mg/l. If the oil and grease is found to be between 0 and 5.0 mg/l, the record will show the result as 5.0 mg/l.

The influent oil and grease analysis is also taken for operational information. As a rule of thumb, if the inlet oil and grease is below 50 mg/l, no operating troubles will be experienced. This plant has never had an oil and grease problem. A single sample is taken three times daily with the exception of the weekend. The method of analysis for the effluent sample is the same as that for the influent.

The running yearly average influent oil and grease content is 30 mg/l. The analyses that exceed the average are to be listed by date. The criteria table of Figure 5.7 sets the information for the Query command.

	A	B	C	D	E	F
1						
2		OIL AND GREASE DATA				
3		(partition-gravimetric method)				
4		(samples in mg/l)				
5						
6		(data table)			(criteria table)	
7	DATE	INFLUENT	EFFLUENT		DATE	INFLUENT
8	1	28	5.0			0
9	2	24	5.0		(output table)	
10	3	30	5.0		DATE	INFLUENT
11	4	35	5.0		4	35
12	5	31	5.0		5	31
13	6		5.0		10	43
14	7		5.1		11	31
15	8	27	7.4		12	33
16	9	27	7.7		16	38
17	10	43	6.7		17	33
18	11	31	5.5		18	48
19	12	33	5.5		20	31
20	13		5.5		24	32
21	14		5.2		30	32
22	15	20	5.0			
23	16	38	7.5			
24	17	33	5.0			
25	18	48	6.9			
26	19	24	5.0			
27	20	31	5.0			
28	21		5.4			
29	22		5.0			
30	23	21	5.5			
31	24	32	5.1			
32	25	21	7.2			
33	26	28	7.3			
34	27		7.7			
35	28		5.6			
36	29	20	6.6			
37	30	32	5.0			
38	31	23	5.7			

FIGURE 5.6. Oil and grease database.

Block	A7..C38
Criteria Table	E7..F8
Output Block	E10..F21

FIGURE 5.7. Query command data.

The first data cell in the field is used to identify the ·location and the formula for the search becomes $+B8 > 30$. Each cell in the field is checked against the formula and if the conditions are met the output is true, if not then false. The 0 in cell F8 merely indicates that cell B8 is below the 30 mg/l average. If cell B8 were above 30 mg/l, the value would be 1.

The output table in Figure 5.6 lists the dates and the analyses that exceed the average. The next step would to be to count those that exceed the average and show them as a percentage of the total. This step will be used in the next chapter on statistics.

6
Statistics

The two sets of statistical functions of Quattro Pro are identical except for the method of designating the block for the calculation. The standard statistics functions set the block by cell location, as in @COUNT(B5···B30). The database statistics functions use the Block, Column, and Criteria as in Table 6.1. Since the database is memory based, the standard statistical functions can be used if the cell locations are known. To have the best of both worlds, either the database or the standard statistical functions can be used on a spreadsheet if the column headings are in a database format.

Database Statistics

Block

The Block for the database or part to be used in the calculation is designated by the upper left and lower right cell coordinates. The Block must include the field names.

Column

The column is a numerical value representing the relative location of the field to be analyzed. The first column in the database is numbered 0. The other columns are numbered sequentially.

Criteria

The Criteria is a cell address containing the criteria information. A cell address might contain the run number, and all data associated with that run could be analyzed. If the assign names command has been used, the criteria expression would be:

Criteria:
Field
+Field > 0

TABLE 6.1. Database statistical functions.

```
@DAVG(Block,Column,      Average values in Column of
      Criteria)         Block that meet Criteria
@DCOUNT(Block,column,    Number of values in Column of
      Criteria)         Block that meet Criteria
@DMAX(Block,Column,      Maximum value in column of
      Criteria)         Block that meet Crtieria
@DMIN(Block,Column,      Minimum values in column of
      Criteria)         Block that meet Criteria
@DSTD(Block,Column,      The population standard dev-
      Criteria)         iation of values in Column of
                        Block that meet Criteria
@DSTDS(Block,Column,     The sample standard of dev-
      Criteria)         iation of values in Column of
                        Block that meet Criteria
@DSUM(Block,Column,      Sum of values in column of
      Criteria)         Block that meet Criteria
@DVAR(Block,Column,      The population variance of
      Criteria)         values in Column of Block
                        that meet Criteria
@DVARS(Block,Column,     The sample variance of values
      Criteria)         in Column of Block that meet
                        Criteria
```

If the uppermost cell of the column is used, the criteria expression would be:

> Criteria:
> Field
> +Upper cell > 0

An example of an statistical command to determine the number of cells with a specific value would be:

$$@DCOUNT\ (A7 \cdots C38,\ 1,\ E43 \cdots E44)$$

where:

	A7 $\cdots$ C38	The database block
	1	The relative column location
	E43 $\cdots$ E44	The criteria data location
	Criteria Data	Field
		+ Field > 0 or + A8 > 0

Using the database statistical functions allows the selection of specific records in the database or spreadsheet. The requirement of the cell location is not needed. Blank cells are ignored but cells containing a label are included as 0.

Spreadsheet Statistics

The spreadsheet statistical functions require the cell locations to identify the List. The List is for either a single column or a block and is typed; for

TABLE 6.2. Spreadsheet statistical functions.

`@AVG(List)`	The average of List
`@COUNT(List)`	A value equal to the number of non-blank cells in List
`@MAX(List)`	The maximim value in List
`@MIN(List)`	The minimum value in List
`@STD(List)`	The population standard deviation of all non-blank values in List
`@STDS(List)`	The sample standard deviation of all non-blank values in List
`@SUM(List)`	The sum of the values in List
`@SUMPRODUCT(Block1,Block2)`	The dot (scalar) product of the vectors corresponding to the blocks
`@VAR(List)`	The population variance of all non-blank values in List
`@VARS(List)`	The sample variance of all non-blank values in List

example, (A1 ··· A35) or (A1 ··· B35). The block must consist of similar data. Blank cells are ignored, but cells with a label are included as 0.

Statistical Methods

Both Figures 6.1 and 6.2 give the calculation of standard deviation and variance for either sample or population statistics. If the all values of a population are available, then the average, standard deviation and variance can be calculated with population statistics where each summation is divided by number of values of the population, n.

If some of the values of a large population are unknown, then an estimate of the variance and standard deviation can be made by choosing a random sample. The sample statistics divides each summation by $n-1$, where n is the number of values in the sample.

Evaluation of Database (Statistics)

Traditionally, the purpose of control of total oil and grease to a sewage treatment plant has been to limit the operational and maintenance problems associated with quantities of petroleum oils. Because of the use of heavy metals as additives, waste petroleum oils and especially lubricating oils are now listed as hazardous materials.

The total oils and grease received consist of both polar and nonpolar materials. The polar compounds are primarily biological lipids and waxes and can include complex aromatic compounds and hydrocarbon derivatives of chlorine, sulfur and nitrogen. The nonpolar materials are aliphatics that comprise most of the lubricating oils. The oil and grease is removed by the organisms in the primary treatment and the activated sludge processes. The heavy metals are not removed by the system but pass through to the effluent and the waste sludge. The treatment plant has always met the heavy metal requirements for both the effluent and the sludge.

As a quick check to determine if a problem did exist, a series of grab samples were taken from the influent. A tentatively approved infrared method with a lower detection limit of 0.2 mg/l was used for the analyses. The polar and nonpolar materials were separated and the amount of each determined for every sample and listed in Figure 6.1. The average percent split between polar and nonpolar compounds was then calculated.

The data from the previous month for total oil and grease, Figure 6.2, were then used to estimate the kilograms/day of lubricating oils entering the plant. The total oil and grease was analyzed by the approved gravimetric method with a lower detection limit of 5.0 mg/l.

The data in Figure 6.1 actually have insufficient samples for a statistical analysis. It also is not random sampling, so it will be treated as samples from a population for the purpose of making a rough estimate. The samples in Figure 6.1 will be used to illustrate the application of the database statistical capabilities of Quattro Pro.

Since Figure 6.1 was to be treated as a database, multiple line headings were not used. The headings are the one word field names as required by a database. The names are assigned using the **Database|Query|Assign Names** command as shown in Figure 5.4. The assigned names are in the block A27 $\cdots$ A30 of Figure 6.1. The criteria are located in the block A27 $\cdots$ E34 of Figure 6.1. There is a criterion for each of the fields, TOTAL, NONPOLAR and POLAR. The calculation of each of the statistical functions is made by identifying the function with the Block, Column, Criteria location. For example, the determination of the TOTAL COUNT is made by:

$$\text{@DCOUNT (B7} \cdots \text{E18, 2, C27} \cdots \text{C28)}$$

where: **@DCOUNT** is the database count function
                B7 $\cdots$ **E18** is the block of data
                2 is the column, when the first column is 0
                C27 $\cdots$ **C28** is the location of the Criteria

The Criteria can be established in the criteria table by using either the assigned name or the location of the first cell in the data field. This is shown in cell D28 as:

$$\text{Either} + \textbf{B7} > 0 \quad \text{or} \quad + \textbf{TOTAL} > 0$$

	A	B	C	D	E
1					
2					
3					
4		INFLUENT OIL AND GREASE			
5		(infrared method)			
6		(samples in mg/l)			
7		DATE	TOTAL	NONPOLAR	POLAR
8		18	38	20	18
9		19			
10		20			
11		21	23	13	10
12		22	31	15	16
13		23	22	14	8
14		24	28	14	14
15		25	21	13	7
16		26			
17		27	19	12	7
18		28	16	11	5
19		COUNT	8.0	8.0	8.0
20		MINIUM	16.0	11.0	5.0
21		MEAN	24.8	14.0	10.6
22		MAXIMUM	38.0	20.0	18.0
23		STD DEV	6.7	2.5	4.5
24		VARIANCE	44.9	6.5	20.0
25					
26	(assign names)			(criteria)	
27	DATE	B7	TOTAL		
28	NONPOLAR	D7		1	(+B7>0 or +TOTAL>0)
29	POLAR	E7			
30	TOTAL	C7	NONPOLAR		
31				1	(+D7>0 or +NONPOLAR>0)
32					
33			POLAR		
34				1	(+E7>0 or +POLAR>0)
35					
36		SELECT VALUES OVER MEAN			
37		(criteria table)			
38		DATE	TOTAL	NONPOLAR	POLAR
39				1	(+NONPOLAR>14)
40					
41		(output table)			
42		DATE	TOTAL	NONPOLAR	POLAR
43		18	38	20	18
44		22	31	15	16
45					
46		NON-POLAR CONTENT = NON-POLAR MEAN/TOTAL MEAN			
47		(14.0/24.8)*100 = 56.5% NON-POLAR			
48		For estimating purposes use 60%			

FIGURE 6.1. Nonpolar oil

	A	B	C	D	E
1					
2					
3			OIL AND GREASE DATA		
4			(partition-gravimetric method)		
5			(samples in mg/l)		
6					
7			(data table)		
8			DATE	INFLUENT	
9			1	28	
10			2	24	
11			3	30	
12			4	35	
13			5	31	
14			6		
15			7		
16			8	27	
17			9	27	
18			10	43	
19			11	31	
20			12	33	
21			13		
22			14		
23			15	20	
24			16	38	
25			17	33	
26			18	48	
27			19	24	
28			20	31	
29			21		
30			22		
31			23	21	
32			24	32	
33			25	21	
34			26	28	
35			27		
36			28		
37			29	20	
38			30	32	
39			31	23	
40					
41			COUNT	23	
42		MINIMUM		20	
43		MEAN		30	
44		MAXIMUM		48	
45		STD DEV		7	
46		VARIANCE		49	

FIGURE 6.2. Total oil and grease.

Each of the database statistical functions is then used with the associated criteria to calculate the data displayed in block C27 ⋯ E24.

As a database, it is possible to display all days of data when the nonpolar content is over the mean of 14.0 mg/l.

The criteria table is located in block A38 ⋯ E39. It should be noted that the block is set up for all the fields, but only the NONPOLAR field has a request. The output table will then display all complete records based upon the request of + NONPOLAR > 14.0 or + D7 > 14.0. The output block is shown in block A41 ⋯ E44 of Figure 6.1.

The monthly influent analyses are in Figure 6.2. This statistical evaluation will use the standard functions as @ Function(List). The standard function does not require a field name type heading, but can use multiple words. The advantage is with a small database or spreadsheet where the cell locations of the data to be used in the calculations are known. There is no need to prepare a criteria table.

The List for each of the functions becomes (C9 ⋯ C39). As an example, the COUNT function would be:

$$@ \text{COUNT (C9} \cdots \text{C39)}$$

The statistical data calculated for the data are shown in block C41 ⋯ C46. The statistical validity of the evaluation is poor because of the limited number of samples. However, it does provide an adequate illustration of the Quattro Pro statistical functions.

Although there was an inadequate number of samples using two different methods of analysis, the mean nonpolar content of the influent would be:

$$10 - \text{E6 liters/day} * 30 \, \text{mg/l} * 60\% = 180 \, \text{kg/day}$$

The estimate of 180 kg/day would also be subject to some unknown variations. But a problem does exist: the source of the lube oil must be determined. In addition, a composite sampling program of the influent must be instituted to provide a valid statistical base. A more accurate approved method of analysis should be negotiated with the toxic board.

7
Iterative Calculations

Quattro Pro provides an iterative method of calculation in the Tools menu, Solve For. By using Solve For, an equation can literally be solved backwards; similar to a trial and error calculation. If a given result is required, the variable that will produce that result can be determined. The variable is adjusted by iteration until the equation balances using the specified result.

Solve For

The command /**Tools**|**Solve For** will bring up the Tools menu in Figure 7.1.

Formula Cell

The /**Tools**|**Solve For**|**Formula Cell** is used to set the location of the formula to be used for the calculation. The formula must be arithmetic and produce a numeric value. The cell can contain a formula that references to formulas in other cells as long as all formulas are consistent and contain a common adjusted variable. Following Quattro Pro syntax, the variables and formula cells can be referred to by name as well as by coordinates.

Target Value

The /**Tools**|**Solve For**|**Target Value** is the result that is wanted from the Formula Cell. The value may be up to plus or minus eight digits. The Target Value may be a cell reference but the Target Value will not change if the reference cell changes. The Target Value cannot be a conditional statement.

Variable Cell

The /**Tools**|**Solve For**|**Variable Cell** is the cell that can be changed to solve for the required value. The Variable Cell cannot be protected or contain a formula, date or text. The Variable Cell must be a contributor to the Formula Cell and also must contain a realistic value.

```
┌────────Tools────────┐
│ Macro               │
│ Reformat            │
├─────────────────────┤
│ Import              │
│ Combine             │
│ Xtract              │
│ Update Links        │
├─────────────────────┤
│ Advance Math        │
│ Parse               │
│ What-If             │
│ Frequency           │
│ Solve For           │
└─────────────────────┘
```

FIGURE 7.1. Tools menu.

```
┌────────Solve For────────┐
│                         │
│ Formula Cell            │
│ Target Value            │
│ Variable Cell           │
│ Parameters              │
│ Go                      │
│ Reset                   │
│ Quit                    │
└─────────────────────────┘
```

FIGURE 7.2. Solve for menu.

```
┌──────────Parameters──────────┐
│                              │
│ Max Iterations        5      │
│ Accuracy          0.005      │
└──────────────────────────────┘
```

FIGURE 7.3. Parameters menu.

Parameters

The command /**Tools|Solve For|Parameters** will bring up the Parameters menu of Figure 7.3. The parameters then can be set for solving the formula of the Formula Cell.

Max Iterations

The number of iterations to be used in the solution is set by the /**Tools|Solve For|Parameters|Max Iterations** command. The default value is 5.

Accuracy

The /**Tools|Solve For|Parameters|Accuracy** command sets the accuracy required for the Target Value. 0.005 is the default value. It should be noted at this point that the equation used in the example caused large value changes with small changes in the Accuracy setting.

Trial and Error Logic

The Solve For command can be simulated in Quattro Pro with the use of the logic functions. The @IF function can be used to determine if a set value has been achieved.

$$@IF(Cond, \ TrueExpr, \ FalseExpr)$$

where: **Cond**........logical expression of the condition
         TrueExpr...value to be used if the expression is true
         FalseExpr...value to be used if the expression is false

Any logical expression to be evaluated as true or false can be used as the formula of Cond. The Cond is evaluated by @IF. If found to be true, the value of TrueExpr is returned. If False, the value of FalseExpr is given.

The use of #AND# or #OR# will allow compound conditions. If #AND# is used, both conditions must be met to be true. The #OR# will require that either one of the conditions must be met to be true.

Vapour Liquid Equilibrium I (Iterative Logic Calculation)

One of the plagues of chemical engineering is the number of trial and error calculations that must be made. The equilibrium conditions of multicomponent vapor liquid systems almost always require the use of trial and error. The computer has simplified and reduced the tedium but not the necessity of iterative calculations.

One of the alternate automative fuels of today is methanol. In this example, methanol has been manufactured by the passage of hydrogen and carbon monoxide over a catalyst. After the passing through the reactor, the methanol is cooled and the pressure reduced. The next step is the removal of the methanol from the by-products by distillation. The feed tray is at 67°C and the still pressure is 760 mm Hg. If the output from the feed heater is 67°C, estimate the percent vaporization of the feed. The basis will be 100 mols of feed with a composition of:

Methanol   81.1% by weight
Ethanol   15.4%   ”   ”
n-Propanol   3.5%   ”   ”

This becomes a simple trial and error calculation.

$$L + V = 100 \tag{1}$$

where: L = total mols of liquid
                                   V = total mols of vapor

By a material balance on each of three components:

$$x_1 L + y_1 V = 100 X_1$$
$$x_2 L + y_2 V = 100 X_2$$
$$x_3 L + y_3 V = 100 X_3 \qquad (2)$$

where: x_1, x_2, x_3 = mol fraction of methanol, ethanol and n-propanol
            in the tray liquid

         y_1, y_2, y_3 = mol fraction of methanol, ethanol and n-propanol
            in the vapor in contact with the residual liquid

     X_1, X_2, X_3 = mol fraction of methanol, ethanol and
            n-propanol in the original liquid before
            vaporization.

By Dalton's and Raoult's laws:

$$y_1 = P_1 x_1 / P$$
$$y_2 = P_2 x_2 / P$$
$$y_3 = P_3 X_3 / P \qquad (3)$$

where: P_1, P_2, P_3 = vapor pressure of methanol, ethanol and
            n-propanol

     P = total pressure.

The sum of the mol fractions of the liquid to the tray after evaporation must equal 1.00:

$$x_1 + x_2 + x_3 = 1. \qquad (4)$$

By combining the above equations, and with the temperature and pressure fixed, the mol fraction of the tray liquid feed can be calculated by trial and error.

The resultant equations are:

$$x_1 = \frac{100\,X_1}{L + P_1/P(100 - L)} \qquad (5)$$

$$x_2 = \frac{100\,X_2}{L + P_2/P(100 - L)} \qquad (6)$$

$$x_3 = \frac{100\,X_3}{L + P_3/P(100 - L)}. \qquad (7)$$

The spreadsheet in Figure 7.4 illustrates the solution to the vaporization problem. Figure 7.5 is a printout of the cell formulas and does not include cells with text. This is what Quattro Pro calls a circular calculation.

The /**Options**|**Recalculation** should be set of Manual until the spreadsheet is prepared; otherwise the spreadsheet becomes blurred with the automatic

```
      A            B          C          D          E        F
 1
 2                  EQUILIBRIUM DISTILLATION
 3                        @67°C
 4
 5 BASIS: 100 MOLS
 6 ----------------------------------------------------------
 7 COMPO-        WEIGHT     MOL        MOLS        MOL        VP
 8   ENT          %        WEIGHT                FRACTION   mmHg
 9 ----------------------------------------------------------
10 METHANOL      81.10     32.04      253.12       0.87      846
11 ETHANOL       15.40     46.07       33.43       0.11      486
12 n-PROPANOL     3.50     60.09        5.82       0.02      218
13 TOTAL        100.00                292.37       1.00
14
15 ----------------------------------------------------------
16 COMPO-     MOL/FRAC   MOL/FRAC    MOLS        MOLS
17   NENT      LIQUID     VAPOR     LIQUID      VAPOR
18 ----------------------------------------------------------
19 METHANOL     0.79       0.88      16.91       69.67
20 ETHANOL      0.16       0.10       3.40        8.04
21 n-PROPANOL   0.05       0.01       0.97        1.03
22 TOTAL        1.00       1.00      21.27       78.73
23 RESIDUAL   21.27192
24 VAPOR               78.73
25              @IF(B22<=1,D22,D22+.01)
```

FIGURE 7.4. Equilibrium distillation.

recalculation. The number of iterations should be set to some number larger than the default of 1. Since the calculations are to continue until the sum of the liquid mol fractions equals 1.0, the iteration number is not critical. A high iteration number will reduce the number of times that F9 has to be pressed. After the entries have been made, F9 is pressed to make the computation.

Note that L, the feed liquid after vaporization, is totalled in D22, and the incremental adjustment is made with B23. B19, B20 and B21 determine the mol fraction of each component in the liquid after vaporization. At equilibrium, the sum of the liquid mol fractions must equal 1.0, B22. The logic formula of B23 is:

$$@IF(B22 < = 1, D22, D22 + .01)$$

The B23 logic formula reads, if the sum of the liquid mol fractions equals 1.0 (true), then B23 becomes D22, the total mols of liquid. If B22 is less than 1.0 (false), D22, the total mols of liquid, are incremented by 0.01 and B22 recalculated.

Since Quattro Pro makes a relative reference when copying formulas, B23 was given an absolute reference by showing it as $B$23. B20 and B21 were then able to be copied from B19. B25 contains a text statement which is identical to B23. The problem can be reset by deleting the value in B23 and copying B25 into B23. Since it is text, the apostrophe must be edited out, then any changes in input can be made and the spreadsheet recalculated with F9.

```
C2:   'EQUILIBRIUM DISTILLATION
D3:   '@67°C
B10:  (F2)  81.1
C10:  (F2)  32.04
D10:  (F2)  100*B10/C10
E10:  (F2)  +D10/$D$13
F10:  (F0)  846
G10:  (F0)  +E10*F10
B11:  (F2)  15.4
C11:  (F2)  46.07
D11:  (F2)  100*B11/C11
E11:  (F2)  +D11/$D$13
F11:  (F0)  486
G11:  (F0)  +E11*F11
B12:  (F2)  3.5
C12:  (F2)  60.09
D12:  (F2)  100*B12/C12
E12:  (F2)  +D12/$D$13
F12:  (F0)  218
G12:  (F0)  +E12*F12
B13:  (F2)  @SUM(B10..B12)
D13:  (F2)  @SUM(D10..D12)
E13:  (F2)  @SUM(E10..E12)
G13:  (F0)  @SUM(G10..G12)
B19:  (F2)  (100*E10)/($B$23+(F10/760)*(100-$B$23))
C19:  (F2)  +F10*B19/760
D19:  (F2)  +B19*$B$23
E19:  (F2)  (100-$D$22)*C19
B20:  (F2)  (100*E11)/($B$23+(F11/760)*(100-$B$23))
C20:  (F2)  +F11*B20/760
D20:  (F2)  +B20*$B$23
E20:  (F2)  (100-$D$22)*C20
B21:  (F2)  (100*E12)/($B$23+(F12/760)*(100-$B$23))
C21:  (F2)  +F12*B21/760
D21:  (F2)  +B21*$B$23
E21:  (F2)  (100-$D$22)*C21
B22:  (F2)  @SUM(B19..B21)
C22:  (F2)  @SUM(C19..C21)
D22:  (F2)  @SUM(D19..D21)
E22:  (F2)  @SUM(E19..E21)
B23:  @IF(B22<=1,D22,D22+0.01)
A24:  [W10] 'VAPOR
B24:  (F2)  100-B23
B25:  '@IF(B22<=1,D22,D22+.01)
```

FIGURE 7.5. Distillation spreadsheet formulas.

```
Formula Cell      B22..B22
Target Value             1
Variable Cell     B23..B23
Parameters
Go
Reset
Quit
```

FIGURE 7.6. Solve for menu.

```
Max Iterations          99
Accuracy             1E-05
```

FIGURE 7.7. Parameter menu.

	A	B	C	D	E	F
1						
2		EQUILIBRIUM DISTILLATION				
3		@67°C using SOLVE FOR				
4						
5	BASIS: 100 MOLS FEED					
6	----	----	----	----	----	----
7	COMPO-	WEIGHT	MOL	MOLS	MOL	VP
8	ENT	%	WEIGHT		FRACT	mmHg
9	----	----	----	----	----	----
10	METHANOL	81.10	32.04	253.12	0.87	846
11	ETHANOL	15.40	46.07	33.43	0.11	486
12	n-PROPANOL	3.50	60.09	5.82	0.02	218
13	TOTAL	100.00		292.37	1.00	
14						
15	----	----	----	----	----	----
16	COMPO-	MOL/FRAC	MOL/FRAC	MOLS	MOLS	
17	NENT	RESIDUAL	VAPOR	LIQUID	VAPOR	
18	----	----	----	----	----	----
19	METHANOL	0.79	0.88	16.91	69.67	
20	ETHANOL	0.16	0.10	3.40	8.04	
21	n-PROPANOL	0.05	0.01	0.97	1.03	
22	TOTAL	1.00	1.00	21.27	78.73	
23	RESIDUAL	21.27189				
24	VAPOR	78.73				
25						
26	----	----	----	----	----	----
27	VARIABLE	MOLS RESIDUAL LIQUID USING ACCURACY				
28	STARTING			OF		
29	VALUE	0.01	0.001	0.0001	1E-05	1E-06
30	----	----	----	----	----	----
31	0	13.365	19.997	21.235	21.272	21.272
32	5	16.424	20.768	21.266	21.266	21.272
33	10	18.803	21.135	21.135	21.272	21.272
34	15	15	20.448	21.256	21.272	21.272
35	20	20	20	21.235	21.272	21.272

FIGURE 7.8. Solve for spreadsheet.

Vapor Liquid Equilibrium II (Solve for Calculation)

The previous example can be solved using the Quattro Pro command /**Tools|Solve For**. The Formula Cell is B22 with B23 as the Variable Cell. The Target Value is 1.0. Figure 7.3 is the Solve For menu with the active cells indicated. The Formula Cell, B22, is the sum of the residual liquid mol fraction calculations of B19, B20 and B21, and by definition must be equal to 1.0. The mol fraction equations are defined in equations 5, 6 and 7 of the previous example. The comman variable for mol fraction is the total mols of liquid, B23.

The Parameter Menu of Figure 7.7 has the Max Iterations set at 99 and the Accuracy set at 1E-05. At the bottom of Figure 7.8 is a table of values of the residual liquid vs the accuracy and starting value used in Solve For. The data indicate a nonlinear relationship of the Variable to the Formula Cell as the Target Value is approached.

The accuracy of the Solve For and the Iterative Logic methods are not comparable. The logical method increments the total liquid mols by 0.01. The calculated sum of the mol fractions is then compared out to 15 significant figures (Quattro Pro Standard) with the Target Value. The Solve For method increments the sum of the mol fractions directly, and requires the increment to be no greater than 1E-05 to obtain the true calculated total mols of liquid.

As with any computer algorithm, care must be exercised in the use of the Solve For command. Any equation that is used with the Formula Cell should be evaluated for the effect of the increment upon the Target Value. Not confirming the accuracy of any algorithm to be used with a design equation could lead to what that famous engineer, Murphy, refers to as "Computer Aided Failure (CAF)."

8
Importing Files

One of the most important features of Quattro Pro is the ability to share files with other programs. Spreadsheets and data files generated with the programs (Table 8.1) can be transferred into the current spreadsheet. Complete or extracted portions of files from Quattro Pro can be exported to these same programs.

Importing Foreign Files

Foreign files with the file name extensions shown in Table 8.1 will be listed in the file directory. The default extension search uses *.W?? and will list all W files. To list all files, including those beginning with other than W, change the prompt to *.*. If a 1-2-3 file with a WK1 extension is brought into Quattro Pro and after being processed, using the /**File|Save** command will save it with the WK1 extension. If the /**File|Save As** command is used, it can be saved with the Quattro Pro WQ1 extension.

Of particular importance in engineering or scientific calculations is reading selected data from a database. When such data are to be imported from a foreign database, the file extension is included for retrieval. The database field names are automatically converted to spreadsheet labels.

It is also possible to import files from programs written in Basic, Pascal, C or other languages. These files are not automatically translated as those with the extensions listed in Table 8.1. The /**Tools|Import** command is used to bring up the Import Menu, Figure 8.1.

The types of files in Figure 8.1 are those that can be imported. Of importance in the transfer of data is the Comma & "" Delimited File. For example, in a Basic sequential file, the data items are separated by commas, and strings are surrounded by quotation marks. A Basic data file can be imported directly into Quattro Pro. The commas cause each data item to be stored in a separate cell as a value. In the case of data pairs, the file will be imported as two columns. The top of the first column will be located at the current position of the cell selector of the receiving spreadsheet. If the data are not separated by commas, Quattro will read each number as text. Data imported as text cannot be used in any further calculations.

TABLE 8.1. Files compatible with Quattro Pro.

Extension	Type of File
.WKQ	Quattro 1.0
.WKZ	Quattro 1.0 (SQZ)
.WKS	1-2-3 1A
.WK$	1-2-3 1A (SQZ)
.WK1	1-2-3 2.01
.WK!	1-2-3 2.01 (SQZ)
.WKE	1-2-3 Educational
.WRK	Symphony 1.2
.WR$	Symphony 1.2 (SQZ)
.WR1	Symphony 2.0
.WR!	Symphony 2.0 (SQZ)
.WKP	Surpass
.RXD	Reflex 1
.R2D	Reflex 2
.DB	Paradox
.DB2	dBase II
.DBF	dBase III,III+ and IV
.DIF	VisiCalc
.SLK	Multiplan

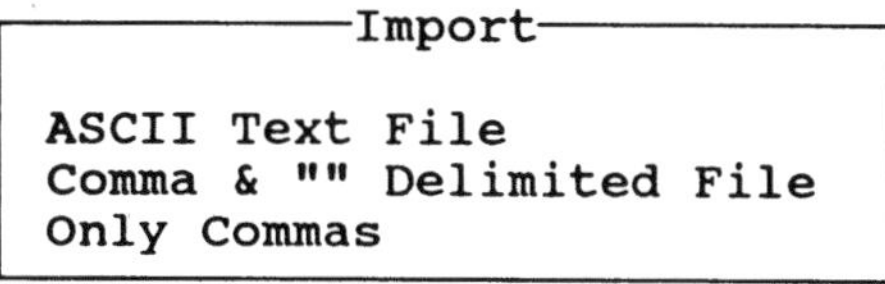

FIGURE 8.1. Import menu.

Exporting Files

Often others will require data from a Quattro Pro spreadsheet or database for use with different programs. Rather than retyping the information, the file can be translated to the required format. Quattro Pro files may be saved to some foreign formats simply by specifying the proper file name extensions as listed in Table 8.1. For example a Quattro Pro spreadsheet could be saved to Lotus 1-2-3 by using the WK1 extension.

Saving a file to a different database format requires that only the fields and data be transferred. Most database formats have a limited space for the field name. To control the block transfer, use the /**Tools**|**Xtract** command, and specify only those cells that contain records and field names. The records can be saved either as values or formulas. The file name and extension are given and Quattro Pro brings up the View Structure Menu, Figure 8.2, for the selected database.

With the View Structure Menu, the layout of the proposed transfer can be viewed in the dBase structure. The field names can then be edited to fit the format and the results entered.

```
dBase - File Save:
View Structure
Write
Quit
```

FIGURE 8.2. View structure menu.

A Quattro Pro spreadsheet or database file may be exported to a word processing file if the program will accept an ASCII text file. Rather than printing the spreadsheet or database file on a printer, the file may be printed to disk. The command /**Print**|**Destination**|**File** will print the file to a disk as an ASCII text file. The text file can be given any suitable name and extension and then imported into the word processing file.

Importing a Basic File

Often data that have been processed by a program written in a language such as Basic, Pascal or C must be used in a Quattro Pro spreadsheet. As an example, data that have been stored as a sequential file with a Turbo Basic program will be imported into a spreadsheet. The values will then be plotted as a graph.

The numeric data in a Turbo Basic sequential file are separated (delimited) by commas. Strings are enclosed in double quotes. The Import Menu of Figure 8.1 of Quattro Pro lists three files, and only the Comma & "" Delimited File can be used to further process data. Files of numeric values imported with either the ASCII Text File or the Only Commas File will become text.

The graph of the cardioid equation $\alpha = D*(1 - \cos \theta)$ is shown in Figure 8.4. The plot is $x = R * \cos(T)$ vs $y = R*S$ in (T). A short Turbo Basic program was used to generate the sequential file:

```
OPEN "CARDIOID.CSV" FOR OUTPUT AS #1        'open file
   PI = 335/113
FOR T = 0 TO PI + 0.1 STEP 0.15
   R = 1 - COS(T)                           'generate function
   PRINT #1, R*COS(T), R*SIN(T)             'write to file
NEXT T
FOR T = 0 TO -PI-0.1 STEP-0.15
   R = 1 - COS(T)                           'generate function
   PRINT #1, R*COS(T), R*SIN(T)             'write to file
NEXT T
CLOSE #1                                    'close file
END
```

Before importing the file, the location on the receiving spreadsheet must be determined for the imported data. The highlighted selector cell will receive the first numeric value and the adjacent cell of that row will recieve the

	A	B	C	D	E
1					
2		TURBO BASIC PROGRAM			
3		CARDIOID			
4		------------------			
5		X	Y		
6		------------------			
7		0	0		
8		0.011103	0.001678		
9		0.042669	0.013199		
10		0.089642	0.043302		
11		0.144157	0.098623		
12		0.19632	0.182891		
13		0.235211	0.296403		
14		0.249994	0.435818		
15		0.231055	0.594307		
16		0.171043	0.762033		
17		0.065734	0.926935		
18		-0.08538	1.075738		
19		-0.27882	1.195108		
20		-0.50721	1.272843		
21		-0.75972	1.298997		
22		-1.02278	1.266838		
23		-1.28114	1.173545		
24		-1.51904	1.020591		
25		-1.72142	0.813762		
26		-1.87514	0.562820		
27		-1.97008	0.280827		
28		-1.99989	-0.01682		
29		0	0		
30		0.011103	-0.00168		
31		0.042669	-0.01320		
32		0.089642	-0.04330		
33		0.144157	-0.09862		
34		0.19632	-0.18289		
35		0.235211	-0.29640		
36		0.249994	-0.43582		
37		0.231055	-0.59431		
38		0.171043	-0.76203		
39		0.065734	-0.92693		
40		-0.08538	-1.07574		
41		-0.27882	-1.19511		
42		-0.50721	-1.27284		
43		-0.75972	-1.29900		
44		-1.02278	-1.26684		
45		-1.28114	-1.17355		
46		-1.51904	-1.02059		
47		-1.72142	-0.81376		
49		-1.87514	-0.56282		
50		-1.97008	-0.28083		
51		-1.99989	0.016815		

FIGURE 8.3. Cardioid data.

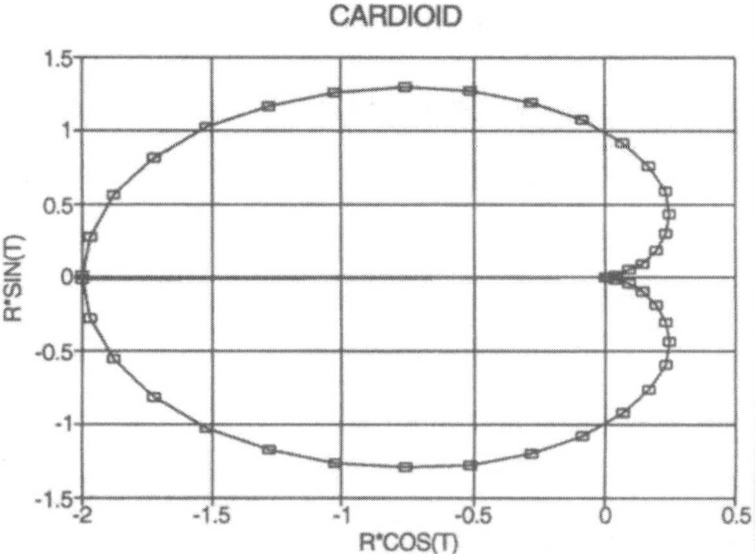

FIGURE 8.4. Cardioid data graph.

second value of the data pair. Each subsequent data pair will be located below the others until all the data have been transferred.

The plot data are stored in file CARDIOID.CSV of the Turbo Basic program file. The command /**Tools|Import|Comma & ''''' Delimited File** will bring up the request for the name of the file to import. After naming the file, the data are transferred to the spreadsheet, as in Figure 8.3.

9
Menu Shortcuts and Macros

At times it would appear that many repetitive keystrokes are needed to access the menus. Quattro Pro allows the creation of shortcuts to access a command with a single key. If repetitive typing is necessary to make up the spreadsheet, then the Macros can be of use.

The menus of Quattro Pro have been presented in a most logical accessible sequence, but some spreadsheets will have different requirements. Many times the menus needed are several layers down in the structure. To reduce menu keystrokes, Quattro Pro can provide shortcuts for the most frequently used menus. All the keystrokes necessary for a specific command can be stored as a combination of CTRL and an alphabetic key.

If the menu change requirements are extensive and are to be permanent, then Quattro Pro allows the redesign of the menu system. The menu items can be renamed and regrouped to suit the conditions by use of Menu Builder. For those few that would require menu changes, Quattro Pro provides an excellent discussion (@ *Functions and Macro, User's Guide*).

The Macro is analogous to the word processing macro that allows the recording and play back of frequently used phrases or layouts. If the design of the spreadsheet is to be simplified or speeded up, Quattro Pro provides for the recording and use of Macros. The Macro allows the storage of multiple keystrokes under a name or a single key.

Menu Shortcuts

To delete a shortcut, locate the menu and highlight the command. Press CTRL-ENTER and press Del twice.

To set up a new shortcut, highlight the desired command in the proper menu. When CTRL-ENTER is held down, the key to be assigned for the command is requested. Hold down CTRL and press the alphabetic key to be used for the shortcut. The CTRL-Key combination is now stored and the Quattro Pro mode becomes Ready. In the interest of remembering a new key assignment, it is suggested that the alphabetic key selected should be a mnemonic device.

TABLE 9.1. Preassigned shortcuts.

Shortcut	Command
Ctrl-A	/Style|Alignment
Ctrl-C	/Edit|Copy
Ctrl-D	/Date Prefix
Ctrl-E	/Edit|Erase Block
Ctrl-F	/Style|Numeric Format
Ctrl-G	/Graph|FAst GRaph
Ctrl-I	/Edit|Insert
Ctrl-M	/Edit|Move
Ctrl-N	/Edit|Search & Replace|Next
Ctrl-P	/Edit|Search & Replace|Previous
Ctrl-R	/Window|Move|Size
Ctrl-S	/File|Save
Ctrl-T	/Window|Tile
Ctrl-W	/Style|Column Width
Ctrl-X	/File|Exit

The Quattro Pro original program has some preassigned shortcuts, as shown in Table 9.1. With the exception of CTRL D, the date prefix, these assigned shortcuts can be reassigned.

Macros

A Macro can be defined as a series of keystrokes or commands stored as labels that can be executed when requested. The macro of Quattro Pro can use any of the keys or commands on the menus, or the special macro commands. The macro functions can extend from user interaction to programming looping, branching and subroutines for automatic execution.

The Macro menu is under the |**Tools heading**. The simplest method of accessing the Macro menu is the shortcut key, Alt-F2. The menu access is /**Tools|Macro** to bring up the Macro menu of Figure 9.2.

Record

The simplest method of producing a macro is to use the Record command of /**Tools|Macro|Record or Alt-F2R**. Quattro Pro records each keystroke and action as the macro is made. The Record command has two options; to record the keystrokes as the menu equivalent commands or as the actual keystrokes. The macro is written as the keystrokes are performed in normal operation. The usual menus, prompts, and messages are displayed. The cell selector is not affected.

To exit the Record mode, select Record again and the Record toggle is set to Off. The exit can also be made from Paste, which asks for a macro

```
┌────────Macro────────┐
│                     │
│ Record              │
│ Paste               │
│ Instant Replay      │
│ Macro Recording     │
├─────────────────────┤
│ Transcript          │
│ Clear Breakpoints   │
│ Debugger            │
│ Name                │
│ Library             │
├─────────────────────┤
│ Execute             │
│ Key Reader          │
└─────────────────────┘
```

FIGURE 9.1. Macro menu.

name and a block in which to paste the macro. The macro is stored and is now available to be used with the Instant Replay command or called up by name. The one caveat of the Record mode is that the typographical errors and the corrections are also recorded. The inexperienced typist may find editing the macro a chore.

Paste

The save the recorded macro, /**Tools**|**Macro**|**Paste** command is used to paste the macro into a spreadsheet. The macro can be accessed whenever the spreadsheet is open.

In addition to using the Record to exit the Record mode, the Paste command can end the macro. The Paste command requests a macro name. The macro can then be executed by name.

The block that will be used to store the macro must be specified. The block does not have to be in the same spreadsheet. A different spreadsheet can be used if the linking rules are followed.

Instant Replay

The last recorded macro can be executed with Instant Play. The /**Tools**|**Macro**|**Instant Replay** command will immediately execute the macro. The Instant Replay can be used until the spreadsheet is exited or another macro is recorded.

Macro Recording

The /**Tools**|**Macro**|**Macro Recording** command provides the method used to record the macro. If Logical is chosen then the macro is recorded as the menu equivalent commands. The Keystroke command will record the actual keystrokes. Keystroke recorded macros are not compatible with all menu trees.

Name

The name of a macro should help to identify the actions. F3 from the input line will bring up the block names choice list which includes the macro names. A macro name cannot be the same as a named block.

The Macro can be named when it is pasted into the cell. The name can be given by /**Tools**|**Macro**|**Name**|**Create** command, and can be assigned with /**Edit**|**Names**|**Create**. The macro name can be changed with /**Tools**|**Name**|**Create** or deleted with /**Tools**|**Macro**|**Name**|**Delete**.

Library

A special spreadsheet may be designated as a macro library. The macro library can be loaded and added to the active spreadsheet. A macro called for execution, if not found on the active spreadsheet, Quattro Pro will then search the open macro library.

The advantage of using a macro library is that the macro does not interfere with the spreadsheet layout. The macros in a library can be inserted into a spreadsheet and any macros developed can be stored in the library for future use.

Using open multiple libraries may be convenient for locating macros. However, if the open libraries have macros of the same name, there is no way to predict which macro will be executed.

A spreadsheet can be designated as a library by using the command /**Tools**|**Macro**|**Library**. Choose Yes from the display menu. A macro can be removed from library status with the command /**Tools**|**Macro**|**Library**|**No**.

Reading Macro Instructions

The default for recording a macro is menu equivalent commands. When the Paste command is used with a recorded macro, the keystrokes for menu selections are translated. The menu equivalent commands, by listing the actions to be taken, provide a description of the functions required. The menu equivalent commands consist of two words preceded by a slash and a space enclosed in curl braces. As an example:

{/ File; Retrieve}

The general action of the command is provided by the first word. The detailed action of the command is specified by the second word. Note that there are no spaces between the first word, the semicolon and the second word.

Editing a Macro

A macro can be changed or modified using the Edit mode with the key F2. Even though the macro is composed of labels, the Edit procedure is identical to a spreadsheet entry. After each line is edited, ENTER is pressed.

In a recorded macro, the key and menu commands are translated into key equivalent and menu equivalent commands. Any changes or additions must be with the same equivalent commands written as labels. To move the cell selector one cell down, the command would be {DOWN}~ or {D}~. To move to a specified cell, the command is {GOTO}A3~ where A3 is the new location. To move a value or a string to a specified location, the command is {/BLOCK; MOVE}A3~. To complete a command, ENTER is required and the tilde~ is used as the symbol.

Writing a Macro with Labels

If a macro is of any length and complexity, reading and debugging become important. Writing the macro as labels allows the use of comments as documentation. Keeping each command to a single cell helps to locate problems. The layout of the macro can improve clarity. For example:

	H	I	J	K	L	M
1	\title	{command}		comment		
2		{command}		comment		

If the above layout is located to the right of the spreadsheet, the title column would be H. The columns I and J would be for the commands, and columns K, L and M would be used for the comments.

Planning a macro is similar to writing a program in a language. If the command macro is complex, preparing a flow chart will provide an accurate guide to the flow.

After moving the cell selector to the beginning block, an apostrophe (') is used to begin the label. With the exception of formulas and @functions that return string values, the Macro must be entered into the cell as a label.

For the use of menu-equivalent commands, Shift-F3 will display a menu of seven macro command categories. Selecting the command includes Quattro Pro writing the command on the spreadsheet input line.

The Quattro Pro *@Functions and Macros* lists for any given menu command the menu-equivalent command. This small manual for the functions and macros is an efficient reference guide.

Debugging

A program to debug a macro is provided by Quattro Pro with the command **/Tools|Macro|Debugger|Yes.** The Debug mode is activated and the status line displays the Debug indicator. A window is opened with the top section containing three rows of the macro. The middle row is the executing macro. The bottom section contains the trace cells for viewing the macro effects on a cell. To bring up the macro to be debugged, the command **/Tools|Macro|Execute** is used. Pressing the SPACEBAR will step through the macro and pressing ENTER will execute the rest of the macro.

From the debug window, the Macro Debugger Commands menu of Figure 9.2 is brought up with FORWARD SLASH (/).

Standard Breakpoints

Breakpoints are a way of executing a macro until a problem with the macro is encountered. Execution pauses at the error. In the Debug window Breakpoints can be selected with FORWARD SLASH (/).

The Block is used to stop execution of the macro at a cell or block of cells. The macro cell is identified by its location, such as E9. Quattro Pro stops operation at the first cell encountered if block is chosen.

If a loop is suspected of being the location of a problem, use Pass Count. The number of passes can be set before operation is stopped. The default number of passes is one, and operation will cease for each pass.

The macro functions at full speed until the break cell is reached. Operation will continue when either SPACEBAR or ENTER is pressed.

```
┌─Macro  Debugger  Commands─┐
│                           │
│       Breakpoints         │
│       Conditional         │
│       Trace Cells         │
│       Abort               │
│       Edit a Cell         │
│       Reset               │
│       Quit                │
│                           │
└───────────────────────────┘
```

FIGURE 9.2. Macro debugger commands menu

```
┌─Breakpoint  Options─┐
│                     │
│     Block           │
│     Pass Count      │
│     Quit            │
└─────────────────────┘
```

FIGURE 9.3. Breakpoint options menu.

```
┌─Conditional─┐
│             │
│   1st Cell  │
│   2nd Cell  │
│   3rd Cell  │
│   4th Cell  │
│   Quit      │
└─────────────┘
```

FIGURE 9.4. Conditional (breakpoint) menu.

```
┌──Macro Trace Cells──┐
│                     │
│  1st Cell           │
│  2nd Cell           │
│  3rd Cell           │
│  4th Cell           │
│  Quit               │
│                     │
└─────────────────────┘
```

FIGURE 9.5. Macro trace cells.

Conditional is used when a logical formula is the breakpoint. The operation stops when the condition is true. Four conditional breakpoints per spreadsheet are allowed. If the results of a macro are to be checked after a given number of steps, the Conditional command can pause at that point. When the counter reaches the set number, the condition is true and operation of the macro stops.

Trace cells can be shown in the Trace window pane. The Trace cell allows monitoring of the contents during debugging. Four trace cell can be viewed at one time.

Reset from the Debugger menu is provided to remove all breakpoints both standard and conditional. Reset will also eliminate the Trace cells. The trace and breakpoint cells can be reset from within the spreadsheet by the use of the command /**Tools|Macro|Clear Breakpoints**.

Quattro Pro provides for editing from within the Debugger Mode. The Edit a Cell command requests the address of the cell and then displays the contents on the input line. The Edit mode functions the same as spreadsheet edit. The modifications are sent to the cell with ENTER.

The Debug mode remains in effect when the macro is finished even though the Debug menu has disappeared. The Debug menu can be brought back with Shift-F2 or /**Tools|Macro|Debugger** and selecting No to exit.

Macro Solve For

As an example of writing a macro by entering the instructions as labels, the Solve For problem of Chapter 7 is repeated. The spreadsheet, Figure 7.8 is used to generate the macro listing in Figure 9.6.

Macro I2 sets the calculation to manual and will start the calculations with the last macro I167. Macros I3 and I4 set the numerical format to two decimal places. Macros I5 to I38 will generate the spreadsheet headings. Input of the compounds, physical properties is done with macros I39 to I61.

The rest of the macros are involved with the calculation using Solve For. Macros I162 to I167 bring in the Solve For program. The Solve For Variable must be a value only and cannot be a formula or a label. The real variable is I149, D22 on the spreadsheet, the sum of the liquid mols. By using macro

BEGIN GROUP

	H	I	J	K	L	M	N
1		Solve For MACRO
2	/M	{/ COMPCALC;MANUAL}~			Set manual calculation
3		{/ BLOCK;FORMAT}~			Set 2 decimal places
4		2~
5		A1..F32~
6		{GOTO}B2~			Generate headings
7		EQUILIBRIUM DISTILLATION~
8		{DOWN}
9		{RIGHT}
10		TEMP = 67°C~
11		{GOTO}A5~
12		BASIS: 100 MOLS~
13		{DOWN}
14		*~
15		{/ BLOCK;COPY}A6~
16		B6..F6~
17		{GOTO}A7~
18		^COMPON-~
19		{DOWN}
20		^NENT~
21		{GOTO}B7~
22		^WEIGHT~
23		{DOWN}
24		^PCT~
25		{GOTO}C7~
26		^MOL~
27		{DOWN}
28		^WEIGHT~
29		{GOTO}D7~
30		^MOLS~
31		{GOTO}E7~
32		^MOL~
33		{DOWN}
34		^FRACTION~
35		{GOTO}A9~
36		*~
37		{/ BLOCK;COPY}A9~
38		B9..F9~
39		{GOTO}A10~			List components
40		{GETLABEL "ENTER CPD:",A10}~
41		{RIGHT}
42		{GETNUMBER "ENTER PCT:",B10}~
43		{RIGHT}
44		{GETNUMBER "ENTER MOL WGT:",C10}~
45		{RIGHT}
46		+B10/C10~
47		{GOTO}A11~
48		{GETLABEL "ENTER CPD:",A11}~
49		{RIGHT}
50		{GETNUMBER "ENTER PCT:",B11}~
51		{RIGHT}
52		{GETNUMBER "ENTER MOL WGT:",C11}~
53		{RIGHT}
54		+B11/C11~			*(Continued)*

```
55          {GOTO}A12~
56          {GETLABEL "ENTER CPD:",A12}~
57          {RIGHT}
58          {GETNUMBER "ENTER PCT:",B12}~
59          {RIGHT}
60          {GETNUMBER "ENTER MOL WGT:",C12}~
61          {RIGHT}
62          +B12/C12~                          Calculate Mol Frac
63          {DOWN}
64          @SUM(D10..D12)~
65          {GOTO}E10~
66          +D10/D13~
67          {DOWN}
68          +D11/D13~
69          {DOWN}
70        +D12/D13~
71          {DOWN}
72          @SUM(E10..E12)~
73          {GOTO}F7~
74          ^VP~
75          {DOWN}
76          ^mmHg~
77          {GOTO}F10~                         Input Vapor Pressure
78          846~
79          {DOWN}
80          468~
81          {DOWN}
82          218~
83          {GOTO}A15~                         Generate Headings
84          \*~
85          {/ BLOCK;COPY}A15~
86          B15..E15~
87          {GOTO}A16~
88          ^COMPO-~
89          {DOWN}
90          ^NENT~
91          {GOTO}B16~
92          MOL/FRAC~
93          {DOWN}
94          ^LIQUID~
95          {GOTO}C16~
96          ^MOL/FRAC~
97          {DOWN}
98          ^VAPOR~
99          {GOTO}D16~
100         ^MOLS~
101         {DOWN}
102         ^LIQUID~
103         {GOTO}E16~
104         ^MOLS~
105         {DOWN}
106         ^VAPOR~
107         {GOTO}A18~
108         \*~
109         {/ BLOCK;COPY}A18~
110         B18..E18~
111         {GOTO}A19~
```

(Continued)

```
112        {/ BLOCK;COPY}A10~
113        A19~
114        {DOWN}
115        {/ BLOCK;COPY}A11~
116        A20~
117        {DOWN}
118        {/ BLOCK;COPY}A12~
119        A21~
120        {DOWN}
121        TOTAL~
122        {DOWN}
123        RESIDUAL~
124        {DOWN}
125        VAPOR~
126        {GOTO}B19~                    Mol/Frac Liquid
127        (100*E10)/(B23+(F10/760)*(100-B23))~
128        {DOWN}
129        (100*E11)/(B23+(F11/760)*(100-B23))~
130        {DOWN}
131        (100*E12)/(B23+(F12/760)*(100-B23))~
132        {DOWN}
133        @SUM(B19..B21)~
134        {GOTO}C19~                    Mol/Frac Vapor
135        +F10*B19/760~
136        {DOWN}
137        +F11*B20/760~
138        {DOWN}
139        +F12*B21/760~
140        {DOWN}
141        @SUM(C19..C21)~
142        {GOTO}D19~                    Mols Liquid
143        +B19*B23~
144        {DOWN}
145        +B20*B23~
146        {DOWN}
147        +B21*B23~
148        {DOWN}
149        @SUM(D19..D21)~
150        {GOTO}E19~                    Mols Vapor
151        (100-D22)*C19~
152        {DOWN}
153        (100-D22)*C20~
154        {DOWN}
155        (100-D22)*C21~
156        {DOWN}
157        @SUM(E19..E21)~
158        {GOTO}B23~
159        0
160        {DOWN}
161        100-B23~
162        {/ MATH;SOLVEFORMULA}B22~    Solve for program
163        {/ MATH;SOLVETARGET}1.00~
164        {/ MATH;SOLVEVARIABLE}B23~
165        {/ MATH;SOLVEMAXIT}99~
166        {/ MATH;SOLVEACCURACY}.00001~
167        {/ MATH;SOLVEGO}
END GROUP
```

FIGURE 9.6. Solve for macro listing

```
          A              B          C          D          E          F

 1
 2                    EQUILIBRIUM DISTILLATION
 3                          TEMP = 67°C
 4
 5       BASIS: 100 MOLS
 6       ********************************************************
 7       COMPON-    WEIGHT      MOL        MOLS       MOL        VP
 8        NENT       PCT       WEIGHT                FRACTION   mmHg
 8       ********************************************************
10       methanol    81.10      32.04       2.53       0.87     846.00
11       ethanol     15.40      46.07       0.33       0.11     486.00
12       propanol     3.50      60.09       0.06       0.02     218.00
13                                          2.92       1.00
14
15       ********************************************************
16        COMPO-   MOL/FRAC  MOL/FRAC    MOLS       MOLS
17         NENT     LIQUID    VAPOR     LIQUID      VAPOR
18       ********************************************************
19       methanol     0.79      0.88      16.91      69.67
20       ethanol      0.16      0.10       3.40       8.04
21       propanol     0.05      0.01       0.97       1.03
22       TOTAL        1.00      1.00      21.27      78.73
23       RESIDUAL    21.27
24       VAPOR       78.73
```

FIGURE 9.7. Macro solve for printout.

I159, spreadsheet cell B23, as the variable in the Solve For Formula, the variable was a value.

Figure 9.7 is the printout of the macro of the Solve For example. Note that cells D22 and B23 have identical values. This is one method of getting around the requirement of not having a formula as a variable.

Index

Made in the USA
Monee, IL
07 July 2026

56550236R00072